Binding and Kinetics for Molecular Biologists

ALSO FROM COLD SPRING HARBOR LABORATORY PRESS

Binding and Kinetics for Molecular Biologists

James A. Goodrich and Jennifer F. Kugel

University of Colorado at Boulder

http://kinetics.cshl.edu

Cold Spring Harbor Laboratory Press
Cold Spring Harbor, New York

Binding and Kinetics for Molecular Biologists

Printed in the United States of America

Publisher	John Inglis
Acquisition Editor	Alexander Gann
Development Director	Jan Argentine
Developmental Editor	Kaaren Janssen
Project Coordinators	Mary Cozza and Maryliz Dickerson
Production Manager	Denise Weiss
Production Editor	Rena Steuer
Desktop Editor	Susan Schaefer
Cover Designer	Ed Atkeson

Library of Congress Cataloging-in-Publication Data

Goodrich, James A. (James Andrew)
Binding and kinetics for molecular biologists / by James A. Goodrich and Jennifer F. Kugel.
p. cm.
Includes bibliographical references.
ISBN-13: 978-087969-736-5 (concealed wire-o : alk. paper)
ISBN-10: 0-87969-736-9 (concealed wire-o : alk. paper)
1. Protein binding. 2. Enzyme kinetics. I. Kugel, Jennifer F. II. Title.

QP517.P76G66 2006
572'.633--dc22

2006018839

10 9 8 7 6 5 4 3 2 1

All Cold Spring Harbor Laboratory Press publications may be ordered directly from Cold Spring Harbor Laboratory Press, 500 Sunnyside Blvd., Woodbury, New York 11797-2924. Phone: 1-800-843-4388 in Continental U.S. and Canada. All other locations: (516) 422-4100. FAX: (516) 422-4097. E-mail: cshpress@cshl.edu. For a complete catalog of all Cold Spring Harbor Laboratory Press publications, visit our World Wide Web Site http://www.cshlpress.com/.

To our families, whose unwavering support and encouragement make every endeavor more fulfilling

Contents

Preface

A COLLEAGUE ONCE COMMENTED THAT "seeing is believing but measuring is knowing." This adage nicely applies to the focus of our book, which covers topics that fall under the general heading of making quantitative measurements of biological binding reactions. Although such topics have been covered in other places, we believe that this book has a specific niche. We strove to present the theories and approaches for quantitating binding reactions using simple, straightforward text that complements the equations. The book is written for individuals at diverse levels of training who desire a resource or teaching tool that explains quantitative measurements of biological binding reactions in a manner that can easily be applied to experimental research, classroom teaching, and/or understanding the work of other scientists. We hope that you find this book to be a valuable resource.

We focus on affinities, kinetics, and regulation of bimolecular binding reactions, which are the fundamental building blocks of all complex biological systems. We begin with an introductory chapter (Chapter 1) that explains the difference between quantitative measurements and qualitative experiments, highlights the importance of measuring the fractional activity of biomolecules, and introduces experimental techniques and principles for monitoring biological binding reactions. Chapter 2 describes approaches to quantitating the affinity with which two molecules interact, assessing binding specificity, and studying competitive binding. Chapter 3 covers cooperativity in three-component systems, where one molecule affects the affinity of interaction between two other molecules, and it includes a discussion of the utility and limitations of Hill plots. Chapter 4 describes measuring forward and reverse rate constants for single-step binding reactions that involve two molecules. Chapter 5 expands on this discussion to explain the kinetics of cooperativity, where one molecule affects the rates at which two other molecules associate and dissociate. Chapter 6 describes how to measure rate constants for a specific type of two-step binding reaction that is found frequently in biology. Finally,

Chapter 7 provides practical information on data analysis, explanations for using nonlinear regression to fit experimental data, and approaches to assessing error in the variables obtained from curve fitting.

The chapters share some common features that are worth emphasizing. Highlighted in grey boxes are "Illustrations" and "Literature Examples," hypothetical experimental scenarios and published experiments, respectively, that exemplify the concepts explained in the chapter. Throughout the book, key equations are contained in boxes and derived in Appendix 1. The important concepts and theories behind the equations are discussed in the chapters, allowing the equations to be used without having to refer to the derivations. Appendix 2, entitled "Resources and References," contains annotated suggestions for further reading and specific resources to expand on the information presented in each chapter. The companion Web site (http://kinetics.cshl.edu) contains computer simulations that can be manipulated to better understand binding and rate curves and fitting data.

Without the important contributions and support of others, this book would not have been possible. We would first like to thank all those at Cold Spring Harbor Laboratory Press who facilitated the creation of this volume from its inception to publication. In particular, Alex Gann initially presented the idea for the book, which we immediately recognized as a useful resource. And Kaaren Janssen's ever-positive attitude and editorial style made the entire process more enjoyable and trouble-free. We would also like to thank the following individuals who read and provided critical comments: Johnny Croy, Rob Kuchta, Pete Mariner, and Dylan Taatjes. We appreciate the patience and understanding exhibited by members of our laboratories when we were preoccupied with writing and editing this book. In addition, we are individually indebted to past mentors and teachers who, often unknowingly, inspired each of us to take a course less traveled in research: kinetic studies of complex biological systems. Finally, and most importantly, we are eternally grateful for our families' encouragement, support, sacrifices, and love.

CHAPTER 1

Performing Quantitative Experiments with Biomolecules

THE IMMEDIATE GOAL OF MANY QUANTITATIVE experiments is to answer the question "how much?" or "how fast?" What concentration of protein is required to bind to a specific DNA sequence? How fast does an RNA dissociate from its target protein? The ultimate goal is to use the answers to such questions to develop models for how biological molecules function in vitro and in cells. Scientists have developed terminology and equations to standardize the discussion and measurement of "how much?" and "how fast?" This book helps navigate the terminology and equations so that they can be readily applied to many biological binding reactions.

> ***Topics covered:***
>
> - Quantitative versus qualitative studies
> - Measuring the concentrations and activities of biomolecules
> - Common techniques used to quantitatively assess binding
> - Experimental considerations and data quantitation

Measuring "how much?" or "how fast?" requires performing quantitative experiments. Therefore, it is important to understand the difference between quantitative and qualitative experiments and when each is most appropriate. One is not superior to the other; each provides unique and potentially useful information. Finally, to perform quantitative experiments, it is essential to have accurate measurements of the total amount and activity of the biomolecules being studied before beginning the experiments. We have provided hypothetical illustrations throughout the chapter to help demonstrate the practical aspects of these topics. In addition, a computer simulation is provided on the Web site (http://kinetics.cshl.edu) that can be manipulated to help visualize concepts presented in the chapter.

TERMS AND PRINCIPLES

Quantitative is formally defined as relating to, concerning, or based on the amount or number of something. When applied to biological binding reactions, quantitative experiments typically seek to place a numerical value on the progress of a reaction or the rate at which it occurs.

Qualitative is formally defined as relating to or based on the quality or character of something, often as opposed to its size or quantity. When applied to biological binding reactions, qualitative experiments typically seek to make observations about a system that often do not involve quantitation of the results.

Concentration versus amount is an important distinction to make when designing quantitative experiments. The amount of a biomolecule is independent of the volume of a solution and can be expressed in units such as grams or moles. Concentration depends both on the amount of the biomolecule and on the volume of the solution and is defined as the amount of a biomolecule in a sample divided by the volume of the sample. The most useful expression of concentration is molar (M; moles/liter).

Standard curves are used to determine the amount of a molecule in a sample. Obtaining a standard curve typically requires titrating a known amount of a biomolecule and measuring the signal intensity for the assay being used. Data points are plotted with signal intensity on the *Y* axis and the amount of the biomolecule on the *X* axis. The points are then fit with a line, and the line is used to quantitate the amount of biomolecule in a sample of unknown concentration.

Linear range describes the region of a standard curve in which the signal intensity increases linearly as the sample amount increases. Amounts of sample that fall outside of the linear range for a given assay are not used in generating the standard curve.

Fractional activity refers to the portion of a sample that is functional. For example, a preparation of an RNA that binds a protein may contain some fraction of molecules that are improperly folded and cannot bind the target protein. The fractional activity of this sample would therefore be less than one, because not all of the RNA will have protein-binding activity.

QUANTITATIVE VERSUS QUALITATIVE

Typically, the ultimate goal of quantitative experiments is not simply to ascertain "how much?" or "how fast?" but to provide insight into how a reaction occurs. Determining quantitative parameters such as affinity between biomolecules and the rates of reaction steps, for example, can provide a model of how a reaction occurs—typically referred to as the mechanism of the reaction. Developing a model for the mechanism of a reaction is qualitative; therefore (ironically), quantitative experiments ultimately provide qualitative descriptions of reactions. For example, consider two proteins that bind overlapping DNA sequences, and both proteins cannot occupy the DNA simultaneously. Quantitative experiments reveal a tenfold difference in the affinities with which each protein binds its respective DNA sequence. Knowing that the concentrations of the two proteins are similar inside cells, one model to describe their function is that the protein with greater affinity typically occupies the DNA and the other does not. Once such a model is developed, further quantitative or qualitative experiments can be designed to test that model.

Determining whether a quantitative or a qualitative experiment is more appropriate depends on the nature of the information sought. If the goal is to determine whether or not a protein binds a specific DNA sequence, a qualitative experiment that simply detects the interaction will suffice. If the goal is to determine the affinity of the protein for a DNA sequence, or differences in the affinities for two different DNA sequences, a quantitative experiment will be required. Often, qualitative experiments provide initial observations that can then be investigated further using quantitative experiments. For example, a qualitative experiment may identify a molecule that inhibits a biological binding reaction. To then characterize the potency of inhibition and/or the mechanism by which the inhibitor functions, quantitative experiments would be required.

Discerning the differences between qualitative and quantitative experiments can help to avoid erroneous conclusions caused by failing to appreciate the limitations of an experiment. In general, qualitative interpretations can be gleaned from quantitative experiments, but the converse is often not true. For example, an experiment that measures the affinity of a protein–protein interaction will provide a quantitative description of the binding energy for the interaction. Finding that two proteins interact with low affinity might also provide the qualitative prediction that other factors likely facilitate this interaction in cells. In contrast, from qualitative experiments, it is typically impossible to draw quantitative interpretations regarding parameters such as affinity, association rates, and kinetic stability. This point is emphasized in Illustration 1.1.

Illustration 1.1. Quantitative versus qualitative western analysis.

Protein A is immunoprecipitated from a cellular extract and associated proteins are detected by western blot with two antibodies: one that detects protein B and one that detects protein C. The band for protein B is much darker than that for protein C in the immunoprecipitated sample (see schematic of gel below). Is more protein B bound than protein C? Unfortunately, a quantitative interpretation cannot be made from the western data. The two antibodies are likely to recognize their epitopes with different affinities, making it impossible to correlate the amounts of proteins B and C in the immunoprecipitated sample with their relative signal strengths. When performed as described, this is a qualitative experiment that leads to the conclusion that both protein B and protein C are in a complex with protein A in the extract.

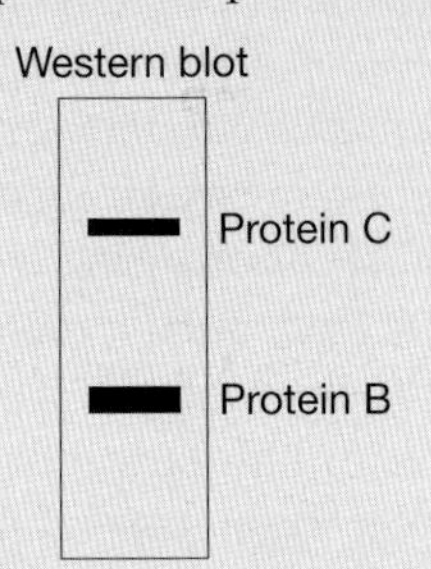

One way to make this experiment quantitative is to include on the western blot titrations of known amounts of proteins B and C in order to generate standard curves. The amounts of protein B and protein C that are coimmunoprecipitated can be calculated using the standard curves and the signal intensities for the experimental bands. This approach is explained in Illustration 1.5.

Even if the experimental result is quantitated, there are limitations on what can be concluded. For example, from the immunoprecipitation results described, it is not possible to conclude that proteins B and C bind directly to protein A; they could associate with a large macromolecular complex that contains A via interactions with other subunits, as illustrated in the schematic below. To measure relative affinities for the two interactions, purified proteins of known concentrations must be used, thereby allowing direct interactions to be studied quantitatively. Techniques to do so are described in Chapter 2.

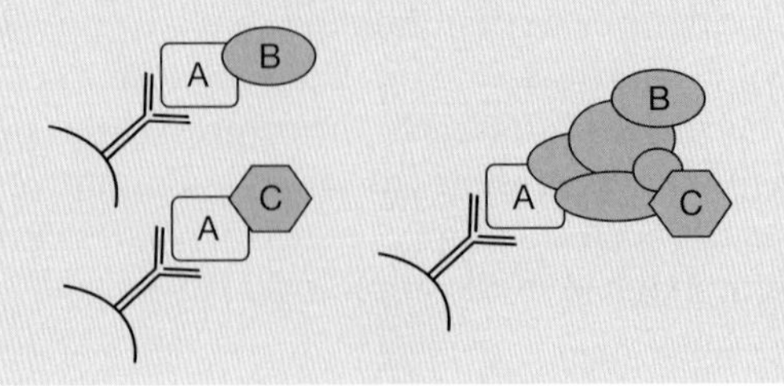

MEASURING THE CONCENTRATIONS AND ACTIVITIES OF BIOMOLECULES

To correctly interpret data from quantitative experiments, it is imperative to know the purity, concentration, and activity of each biomolecule in the experiment. For example, a measurement of the affinity of an interaction between two biomolecules is only as good as the measurements of the concentrations of each of the interacting partners. In the following sections, we discuss general considerations for determining purities, concentrations, and activities of biomolecules.

Sample Purity

It is important to consider the purity of a sample before measuring the concentration of a biomolecule in the sample. For example, contaminating cellular nucleic acids in a plasmid prepared from bacterial cells will increase the UV absorbance of the sample, thereby artificially inflating the apparent concentration of the plasmid. If a sample is not pure, then either additional purification must be performed or a method must be used to determine the concentration of only the biomolecule of interest in the nonhomogeneous sample, usually by viewing that molecule directly, as opposed to quantitating the total amount of molecule(s) in the sample. For example, proteins and nucleic acids can be separated by gel electrophoresis, thereby allowing the desired biomolecule to be viewed independently of other contaminating biomolecules in the sample. Quantitative experiments can be performed with a nonhomogeneous sample if two conditions are met: (1) Techniques are employed to quantitate only the desired molecule and not contaminants when determining concentration and (2) the contaminants do not influence the reaction to be studied.

Measuring Concentrations of Biomolecules

Performing accurate quantitative experiments usually requires determining the concentrations of biomolecules in stock solutions; for example, a plasmid DNA preparation, a newly prepared recombinant protein, or a small molecule ligand. Techniques to measure concentration rely on the comparison of a sample of unknown concentration to a linear standard curve. For certain instruments, the researcher need not generate this curve because others have predetermined the linear range of detection. For example, using a spectrophotometer to determine the concentration of DNA in a solution does not require the generation of a standard curve because it has been determined that absorbance (A_{260}) measurements are linear between 0.2 and 1.0. Typically, the DNA sample is diluted so that the A_{260} of the diluted sample is between 0.2 and 1.0 and the concentration of DNA is calculated from the A_{260}, the extinction coefficient for DNA (which is known), and the dilution factor. This same concept applies to determining the concentration of a pure protein; the extinction coefficient of the protein and the absorbance at 280 nm can be used to calculate the concentration. Alternatively, any of several commercially available reagents (e.g., a Bradford assay) can be used.

In contrast, many techniques to determine the concentrations of biomolecules will require generating a standard curve. For example, if a DNA or protein sample is too dilute, too impure, or in too small a volume to determine its concentration by measuring the absorbance, then a standard curve must be generated. Gel electrophoresis can be used to determine the concentration of the DNA or protein in the sample as follows. A sample of a known amount of DNA or protein is run on the appropriate gel with the unknown sample to generate a standard curve. The gel can be quantitated and the concentration of the unknown sample determined from the standard curve, as explained in Illustration 1.2.

Illustration 1.2. Generating a standard curve and determining the concentration of a protein.

Protein K is a recombinant protein that has been expressed in *Escherichia coli* and purified by chromatography. A contaminating *E. coli* protein is present in the preparation of protein K as determined by SDS-PAGE (denaturing polyacrylamide gel electrophoresis). How should the concentration of protein K be determined? A technique such as absorbance at 280 nm or a Bradford assay cannot be used because the contaminating protein would contribute to the signal. Instead, protein K should be resolved from the contaminating protein by SDS-PAGE and then visualized and quantitated using a standard curve, as described below.

Protein K and a titration of known amounts of bovine serum albumin (BSA) (or another convenient protein) are subjected to electrophoresis through the same gel, as shown below.

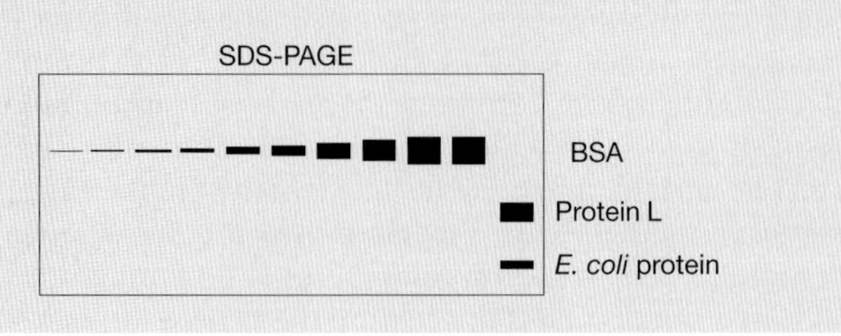

The gel is stained with Coomassie, the intensity of each BSA band and protein K band is quantitated by densitometry, and the BSA results are used to generate a standard curve. A plot is generated with the signal intensity of the BSA bands on the *Y* axis and the amount of BSA on the *X* axis (circles in the plot). The points in the region of the curve that is linear are fit with a line to generate a standard curve that relates the amount of protein to a signal on the gel. (Fitting data with a line is discussed in Chapter 7.) Notice that some points at both the low and high amounts of BSA are not included in the curve fit, because, for those amounts of protein, the signals are out of the linear range.

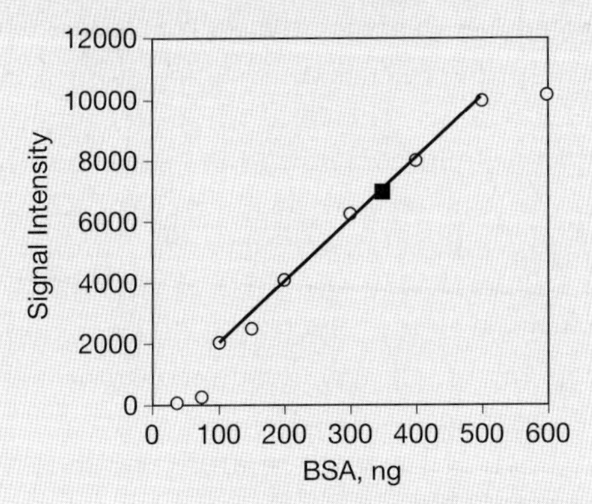

The solid square on the plot is the signal from the protein K band, which falls within the linear range of the standard curve. Importantly, if the signal intensity for this band were outside of the linear range, then this gel could not be used for quantitation of the amount of protein K. Another gel would have to be run, including known amounts of BSA to generate the standard curve and an amount of protein K that falls within the linear range. Using the equation for the line and the signal from protein K for the *Y* value, the amount of protein K that was loaded on the gel can be calculated.

Sample calculation to determine amount of protein K:

Equation that fits the standard curve: $Y = 20X - 84$

Signal intensity from protein K sample (filled square): 6921

Using this value as *Y*, solve for *X* to obtain 350 ng of protein K

The molecular mass of protein K is 50 kD; therefore, 7 pmoles of the protein was loaded on the gel. The volume of the protein sample loaded on the gel was 5 μl; therefore, the concentration of protein K in the sample is 1.4 μM.

Assessing the Activity of Biomolecules

After the purity and the concentration of a sample are assessed, the activity of the biomolecule must be determined. Occasionally, preparations of biomolecules consist of mixtures of active and inactive forms, for reasons that may not be known. The portion of the total biomolecule that has activity is referred to as the fractional activity. It is important to measure the fractional activity of a sample before embarking on quantitative experiments for at least two reasons. First, if the majority of biomolecules in a sample are inactive, then the sample may need to be further purified or repurified to remove inactive molecules that could potentially inhibit the function of the active molecules. Often, it can be helpful to assess fractional activity at steps during the purification process. Second, knowing the fractional activity allows the concentration of the active species in solution (rather than the total concentration) to be used in quantitative studies. For example, a protein that has been purified to homogeneity may consist of some molecules that are active and some that are inactive (e.g., misfolded). If the total concentration of protein in the sample is used to design and perform quantitative experiments, then the measurements made will be incorrect. Clearly, the concentration of active protein in the sample should be used.

Determining the fractional activity of a preparation of a biomolecule requires using an activity assay for the reaction of interest. The simplest way to describe how to determine fractional activity is through Illustrations. In Illustrations 1.3 and 1.4, we discuss two approaches for determin-

Illustration 1.3. Measuring the fractional activity of an RNA that binds a protein.

A new preparation of a ^{32}P-labeled small noncoding RNA is made, and the fractional activity of the preparation must be determined. The protein known to bind this RNA is available, and the RNA–protein complex can be separated from free RNA using native gel electrophoresis. A series of binding reactions is set up with RNA held at a constant concentration and the amount of protein is titrated. The RNA–protein complex in each reaction is separated from free RNA using native gel electrophoresis, and phosphorimagery is used to quantitate the amount of bound RNA and free RNA in each case. The fraction of RNA in a complex with protein is calculated (bound/[bound + free]) and plotted versus the concentration of protein added to reactions, as shown in the plot on page 9 (the data points are simply connected with lines). The fraction bound reaches a plateau at a value of 0.85; hence, the fractional activity of the RNA preparation is 0.85. In other words, 85% of the RNA is capable of binding the protein. The other 15% of the RNA cannot bind the protein; perhaps this RNA does not have the native secondary or tertiary structure. Therefore, the concentration of functional RNA, which is used in quantitative experiments, is 85% of the measured concentration of total RNA.

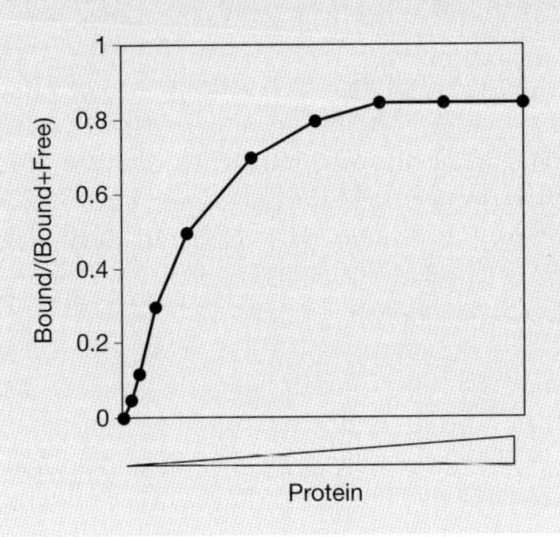

ing the fractional activity of a molecule. Because binding reactions are the primary focus of this book, these reactions are used as examples. First, we discuss how to determine fractional activity for a molecule that can readily be labeled and detected.

In some cases, it might not be possible or practical to directly visualize (e.g., label) the molecule for which fractional activity needs to be measured. This is often the case when the molecule of interest is a protein. It is possible, however, to measure the fractional activity of an unlabeled molecule by observing its interaction with a labeled binding partner, as explained in Illustration 1.4 (see also Simulation S1 on the Web site). Note that it is critical to know the stoichiometry of the molecules in the complex and to use the concentration of the functional oligomer. For example, if a protein binds a target as a dimer, the concentration of dimer (i.e., half of the concentration of monomer) should be used in reactions.

Illustration 1.4. Measuring fractional activity of a DNA-binding protein.

A new preparation of protein A, which binds DNA as a dimer, has been made and the fractional activity for binding its target DNA must be determined. Protein A cannot be easily labeled, so a DNA fragment containing the binding site for protein A is ^{32}P-labeled to allow monitoring of the DNA–protein A complex by filter binding. The approach to measure fractional activity requires two steps. First, the amount of DNA required to fully saturate a set amount of protein A is determined. (This step is required because in the second step, the concentrations of both the protein and the DNA need to be above the affinity constant that governs their interaction. For a detailed discussion of this concept, see Chapter 2.) A series of reactions is set up in which the concentration of protein A dimers is held constant at 10 nM, and the DNA is titrated from 10 nM to 800 nM. The DNA–protein A complex that forms is monitored by filter binding, followed by scintillation counting to determine the amount of complex retained on the filter. As the DNA is titrated up, the amount of complex plateaus when the protein A dimers are fully saturated. A concentration of 100 nM DNA is well within this plateau region, so this concentration of DNA is chosen for the second step.

A second series of reactions is set up in which the DNA is now held constant at the saturating amount (100 nM as determined in the first step), and protein A dimers are titrated from 10 nM to 500 nM. As in the first step, the complex is monitored using filter binding. Scintillation counts of complexes retained on the filters are plotted versus the concentration of protein A dimers added to reactions, as shown below. The plot indicates, given a DNA concentration of 100 nM, that a concentration of 140 nM protein A dimers is required to produce the maximum amount of complex. Dividing the concentration of DNA used in the reactions by the concentration of protein A dimers at saturation gives the fractional activity of the protein A preparation, which is calculated to be 0.7 (100 nM/140 nM). Thus, the concentration of functional dimers in subsequent experiments would always be equal to 70% of the total concentration of protein A.

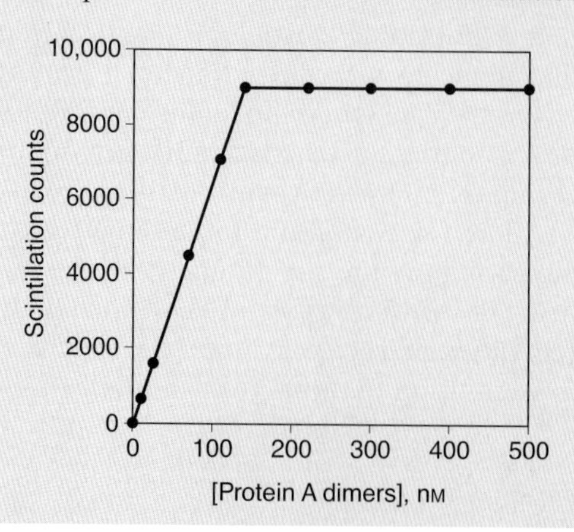

COMMON TECHNIQUES USED TO QUANTITATIVELY ASSESS BINDING REACTIONS

To study binding reactions quantitatively, it is critical to establish an appropriate and reliable assay. This involves developing a means to monitor the complex separately from the free biomolecules. The potential methods to do so are numerous; however, we discuss some of the most commonly used methods here. These are grouped into three general categories; for a given binding reaction, one or more of the methods will often be applicable. The following descriptions are meant to introduce the assays and to provide advantages and disadvantages for each; they are not descriptions of how to perform the techniques.

Assays That Separate Complexes from a Solution

Many of the most commonly used techniques to monitor binding between biomolecules involve the separation of complexes from free biomolecules. In these methods, complexes are allowed to form in solution and then the solution is fractioned, often relying on the separation of molecules based on differences in size, charge, or conformation.

Filter-binding assays. Filter-binding assays can be used to study protein–DNA and protein–RNA interactions. DNA and RNA do not bind nitrocellulose, whereas proteins do. When a solution containing a protein–DNA or protein–RNA complex is filtered through nitrocellulose, the protein–nucleic acid complex will bind the filter (as will free protein), but free DNA or RNA will pass through the filter. Typically, the DNA or RNA used in the binding reaction is ^{32}P-labeled, and the amount of DNA or RNA retained on a filter is quantitated by scintillation counting to assess the amount of complex.

ADVANTAGES

- Many samples can be analyzed with ease and rapidity.
- There is a high sensitivity for detecting bound nucleic acid.
- The method provides a high level of accuracy and reproducibility.

DISADVANTAGES

- It is not possible to distinguish between specific and nonspecific protein–nucleic acid complexes.
- It is not possible to separate different protein–nucleic acid complexes that are present in a single solution.

Cell association assays. This is a general term for a variety of experiments that measure the binding of molecules (often small molecule ligands) to cells. A labeled molecule is incubated with cells, and the cells are subsequently isolated from the solution by filtration or centrifugation. The amount of bound molecule is then quantitated (typically by scintillation counting or fluorescence). The advantages and disadvantages are very similar to those for filter-binding assays.

Gel-filtration chromatography. This method separates molecules based on their size and shape by sieving them through a matrix. It can be used for complexes containing protein, DNA, RNA, and small molecules. Gel filtration can involve large columns taking hours to run or small spin-columns taking minutes to run using a microcentrifuge.

ADVANTAGES

- Mixtures of all types of free and bound biomolecules (including different oligomeric complexes) can be separated and analyzed.
- None of the biomolecules to be analyzed need to be labeled.
- Information on the size of globular complexes can be obtained.
- Gel filtration can be performed under a variety of buffer conditions and temperatures.

DISADVANTAGES

- It is not practical for most kinetic experiments, given the time that it takes to run the column.
- Some complexes might not be stable enough to survive the chromatography.
- It can be difficult to perform chromatography on multiple samples.

Electrophoretic mobility shift assays (EMSAs). In this method, complexes are separated from free biomolecules by electrophoresis through a native gel (typically polyacrylamide). For a given set of electrophoresis conditions (e.g., electric current, buffer composition, temperature, and gel percentage), the rate at which molecules migrate through the gel matrix depends on molecular weight, charge, and shape. The most common use of EMSAs is for the separation of protein–DNA complexes from free DNA. The complex migrates slower than free DNA through the gel, allowing both bound and free DNAs to be observed as distinct bands and quantitated.

ADVANTAGES

- Labeled molecules are detected with a high degree of sensitivity.
- Complexes containing only a fraction of the labeled molecules can be detected.
- The assay is technically straightforward to perform (solutions containing biomolecules are simply loaded on the gel).
- The gel can often resolve multiple different complexes and the free biomolecule in a single lane.

DISDVANTAGES

- Not all complexes that form in solution survive the electrophoresis conditions.
- Kinetic experiments are often not feasible unless the rates being investigated are very slow.
- The oligomeric state of complexes cannot be directly determined or inferred from the position of migration in the gel.

Assays that Detect Complexes in Solution

These assays allow binding reactions to be directly monitored in solution. This could be via assessing changes in the solution in which the reaction occurs (e.g., temperature, absorbance, and light emission) or by probing changes in one of the biomolecules with a secondary solution assay (e.g., protection against an enzyme that cleaves one of the interacting biomolecules or protection against a chemical that modifies one of the interacting biomolecules). Two examples are discussed below.

Fluorescence. A number of fluorescence techniques (e.g., fluorescence resonance energy transfer, fluorescence anisotropy, induced fluorescence, and fluorescence quenching) allow binding to be studied by monitoring an increase or decrease in fluorescence due to association of two biomolecules. Often, one or both of the biomolecules are labeled by the addition of a fluorescent dye. Alternatively, it is sometimes possible to monitor changes in a natural fluorophore present in a biomolecule (e.g., a tryptophan in a protein).

ADVANTAGES

- It is possible to study the binding of molecules without perturbing the solution.
- It is possible to measure rate constants for very rapid binding reactions (i.e., the reaction can be monitored in real time).
- Many samples can be analyzed with ease and rapidity.

DISADVANTAGES

- The addition of fluorescent dyes to biomolecules could affect their function.

Protection assays. A second method for detecting complexes in solution relies on the fact that formation of the complex buries surfaces on the interacting molecules, thereby protecting these surfaces from reacting with other reagents added to the solution, such as cleavage enzymes or chemical modifiers. Commonly used enzymatic protection assays include DNase or RNase footprinting, in which a region of a DNA or RNA is protected from a nuclease, and protease protection, in which a region of a protein is protected from protease cleavage. Examples of chemical protection assays include methylation protection, in which a region of a DNA or RNA sequence is protected from methylation by a bound molecule, and deuterium exchange, in which binding of another biomolecule protects a region of a protein from exchanging deuterium and hydrogen in amides. The enzymatic or chemical modification itself is performed quickly and provides a "snapshot" of the complexes in solution at the time of addition of the enzyme or chemical.

ADVANTAGES

- Complexes may be probed while in solution.
- Typically, very little material is required.
- The secondary assays themselves typically take seconds; hence, the exposure time for taking the snapshot is short.

DISADVANTAGES

- The reagents themselves (especially those involved in chemical modification) can directly influence the properties of the biomolecules in the complex.
- These assays require that minimally 10% of the molecule being modified be bound by the other biomolecule(s) in the reaction; however, in reality a much higher percentage will likely be needed.

Assays in which a Biomolecule Is Immobilized

Some binding assays utilize biomolecules that are immobilized on the surfaces of beads, 96-well plates, or glass coverslips as a means to separate complexes from free molecules.

Affinity resins. Biomolecules can be immobilized on many different commercially available resins (e.g., streptavadin agarose beads, protein A–Sepharose with a bound antibody, or *N*-hydroxy-succinimide agarose). After incubation with a solution containing a potential binding partner, the resin is separated from the solution (e.g., by centrifugation) and washed extensively with an appropriate buffer, and the amount of bound biomolecule is quantitated.

ADVANTAGES

- A single affinity resin can be used to investigate the binding of many potential partners.
- Reactions (including washes) can be performed under many buffer conditions.
- The amounts of both bound and unbound molecules (those remaining in solution after resin is isolated) can be quantitated.

DISADVANTAGES

- The resin itself might have some affinity for the molecule in solution.
- Immobilization of a molecule might affect its binding activity.
- It may be difficult to perform kinetic experiments because washing the resin after binding can be relatively slow.

Surface plasmon resonance. This method utilizes a special machine (e.g., a Biacore) that detects changes in the refractive index of a gold-coated glass slide due to changes in the mass of the aqueous layer close to the surface. To perform the experiments, one molecule is first immobilized on the surface. A solution containing the interacting partner is flowed over the surface using a microfluidics system. The system detects the association of molecules from the solution with the immobilized molecules, which causes a change in mass in the aqueous layer near the sensor surface.

ADVANTAGES

- The method provides the ability to collect real-time kinetic data for association and dissociation reactions.
- Labeling of molecules is not necessary.
- Only small amounts of sample are needed.

DISADVANTAGES

- The surface can influence the binding reaction.
- Directly measuring affinities between molecules is difficult.
- The approach requires an expensive and highly specialized instrument.
- Often, a substantial amount of time is required to develop reliable methods for each interaction to be studied.

EXPERIMENTAL CONSIDERATIONS AND DATA QUANTITATION

We have highlighted below a few key points to consider when designing quantitative experiments involving biological reactions. This list is by no means all-inclusive, but it illustrates that, in general, accurate quantitative experiments must be carefully designed.

Assay Conditions

Most biological reactions are performed in a buffered solution under conditions that mimic physiological conditions. The components of the solution can greatly affect the outcome of the experiment. In other words, specific parameters that describe the nature of a reaction such as an affinity constant or a rate constant will change with the buffer conditions. When designing an experiment, consider that many conditions have the potential to affect the values of quantitative measurements: pH, salt concentration, concentration of divalent cations (e.g., Mg^{++}), and temperature.

Accounting for Background in Measurements

One of the most important aspects of quantitating data from biological binding reactions is that the background inherent to the assay must be quantitated and subtracted from the data. For example, in filter-binding assays, a small amount of ^{32}P-labeled nucleic acid is typically retained on the filter even in the absence of added protein. It is important to perform a control reaction to measure this amount of radioactivity and to subtract this value from each of the experimental data points. Similarly, in EMSA experiments, which are often quantitated using phosphorimagery, it is important to quantitate the background level of phosphorimager units present in a region of the gel that does not contain a specific band. Moreover, the level of background often varies up and down, as well as across the gel. For this reason, it is best to obtain a measurement of background near each band that will be quantitated; hence, a unique background value is subtracted from the signal for each band. The background-corrected data are used for data analysis.

Data Points and Plots

Many quantitative experiments ultimately involve plotting data points and fitting the points with the appropriate equation for the measurement being made. Plots fall into two general categories: lines and curves. In each case, a minimum number of data points is needed for the measurements to be accurate. In general, at least 5 points are needed to establish the slope and *Y* intercept of a line and at least 12 points are needed to establish the shape and plateau of curves. As the number of points increases, the parameters obtained from fitting the data become more reliable. Note, however, that in some situations, it might not be feasible to obtain enough points to meet these guidelines. Quantitative experiments can still be performed; however, the results should be interpreted conservatively.

Illustration 1.5. Quantitating the amount of a specific protein in a coimmunoprecipitation.

Recall the experiment described in Illustration 1.1 in which it was not possible to determine whether more of protein B or C coimmunoprecipitated with protein A. Repeating the western blot and quantitating it in the following manner will provide the desired information. Including titrations of known amounts of proteins B and C on the gel with the immunoprecipitated sample will reveal which protein is present in a greater amount in that sample. The western blot is visualized by chemiluminescence, and the light emitted from each band in the titration is quantitated and corrected for background. As shown in the plot below, the signal from each band in the titration is plotted versus the amount of protein, and the points that fall in the linear range for each data set are fit to a line, thereby generating two standard curves: one for protein B (open squares) and one for protein C (open circles).

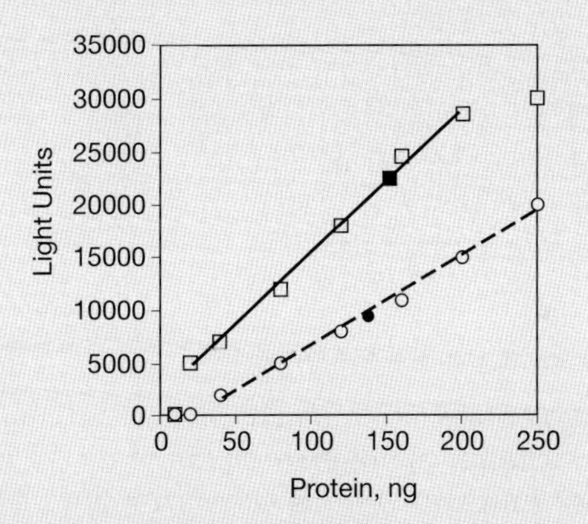

The filled square and filled circle are the signal intensities of the immunoprecipitated sample for proteins B and C, respectively. The signal intensities for both of the experimental bands (i.e., the coimmunoprecipitated proteins) fall within the linear range of the appropriate standard curve. Therefore, we can calculate the amount of protein B or protein C that coimmunoprecipitated using the equation for a line as illustrated in the sample calculation below. Importantly, if the signal intensity for either protein band were outside of the linear response range, or if the standard curve itself were nonlinear, quantitation with this western would not be possible.

(Continued on following page)

Illustration 1.5. *(Continued)*

Sample calculation to determine amount of B:

Equation that fits the solid line: $Y = 136X + 1806$

Light units from protein B sample (filled square): 22441

Using this value as Y, solve for X to obtain 152 ng of protein B

Given the molecular mass of protein B (30 kD): 5 pmoles

Applying this same calculation to sample C and the equation for the dashed line, we can determine that 136 ng of protein C were present in the coimmunoprecipitated sample, which, given the molecular mass of protein C (70 kD), is equivalent to 2 pmoles. Therefore, more moles of protein B than protein C coimmunoprecipitated with protein A.

CHAPTER 2

Affinity Constants

AFFINITY CONSTANTS are numeric representations of the strength with which two molecules interact. Therefore, affinity constants provide quantitative meaning to phrases such as "tight binding" and "weak interaction." Beyond placing quantitative values on the strength of interactions, affinity constants can provide insight into the mechanisms of interactions and, when coupled with biological experiments, can aid in predicting if and when specific interactions function in cells.

The intent of this chapter is to provide an understanding of affinity constants and how to experimentally measure them for bimolecular interactions. The information presented here is built upon in Chapter 3 to explain affinity and cooperativity in higher-order complexes. Throughout the chapter, hypothetical examples from the illustrations and examples from the literature are provided to demonstrate practical aspects of measuring affinity constants, as well as to convey how affinity constants can provide a greater understanding of biological systems. Key equations are contained in boxes and derivations are provided in Appendix 1. In addition, computer simulations are provided on the Web site (http://kinetics.cshl.edu) that can be manipulated to help visualize concepts presented in the chapter.

Topics covered:

- Affinity constants for bimolecular interactions
- Competitive binding (IC_{50})
- Binding specificity
- Homodimerization
- Using enzyme reactions to measure affinity constants

TERMS AND PRINCIPLES

Equilibrium is the state that is attained when all processes affecting a reaction are in balance. With respect to interactions, equilibrium exists when there is no overall change in the concentrations of the free and bound species in the reaction over time. Importantly, at equilibrium, a binding reaction is not static. Within the population, complexes are constantly forming while others are dissociating; however, there is no net change in bound versus free.

Affinity describes the strength of an interaction between two molecules.

***Dissociation constants* (K_D) *and binding constants* (K_B)** are quantitative parameters describing affinity that can be experimentally measured. For the sake of clarity, we will use the term "affinity constant" to refer in general to both equilibrium dissociation constants (K_D) and binding constants (K_B). Affinity constants are measured under equilibrium conditions. A K_D is expressed in units of m (moles/liter) and a K_B is expressed in units of M^{-1}, making them reciprocals of each other. K_D is inversely related to affinity; as affinity increases, K_D decreases. K_B is directly related to affinity; as affinity increases, K_B increases.

Bimolecular interactions occur very frequently in biology and can be represented as A + B ↔ AB. Interactions of this type are the easiest to understand and to study quantitatively. Approaches to measuring affinity constants for bimolecular interactions are the focus of this chapter.

A competitive inhibitor binds to a molecule and blocks its association with a partner molecule. For example, a competitive inhibitor (X) of the bimolecular interaction A + B ↔ AB blocks the formation of AB by binding to A at the same site to which B normally binds. Therefore, in the presence of X, both AX and AB can form. The concentrations of X and B and their affinities for binding A will determine the ratio of the two complexes in the system.

IC_{50} is the concentration of a competitive inhibitor that blocks the formation of a bimolecular complex by 50%. Using the example from the previous paragraph, the IC_{50} would be the concentration of X required to decrease the amount of the AB complex by 50% at constant concentrations of A and B.

Specificity describes the relative affinities with which a molecule interacts with different partner molecules. For example, a protein binds a DNA sequence with high specificity only if it associates weakly with other DNA sequences.

BIMOLECULAR INTERACTIONS (A + B ↔ AB)

The most fundamental interaction in biology is the bimolecular binding reaction. It is the building block from which all higher-order interactions are assembled. Therefore, knowledge of the affinity constants governing bimolecular interactions enhances our understanding of biological regulation. We now discuss in detail how to measure affinity constants for bimolecular interactions.

Theory and Equations

Bimolecular binding interactions can be described by:

$$\mathrm{A} + \mathrm{B} \underset{k_{-1}}{\overset{k_1}{\rightleftharpoons}} \mathrm{AB} \tag{1}$$

where A and B represent two different interacting biomolecules.

Relating affinity constants to rate constants. To understand affinity constants, it is helpful to consider the rate constants governing the association and dissociation of the two molecules (also see Fig. 2.1). (For detailed discussions of rate constants and approaches to measuring them, see Chapter 4.) The top arrow in Equation 1 indicates the association of A and B, which is governed by a forward rate constant (k_1) with the units of $\mathrm{M}^{-1}\mathrm{s}^{-1}$. The bottom arrow indicates the dissociation of the AB complex, which is governed by a reverse rate constant (k_{-1}) with the units of s^{-1}. For this discussion, we presume that the reaction is at equilibrium and there are no net changes in the concentrations of A, B, and AB.

Affinity constants relate directly to the forward and reverse rate constants that govern a reaction. The K_D is equal to the reverse rate constant divided by the forward rate constant.

$$K_D = \frac{k_{-1}}{k_1} \qquad \text{units: } \mathrm{M} = \frac{\mathrm{s}^{-1}}{\mathrm{M}^{-1}\mathrm{s}^{-1}} \tag{2}$$

The K_B is equal to the forward rate constant divided by the reverse rate constant.

$$\mathrm{A} + \mathrm{B} \underset{k_{-1}}{\overset{k_1}{\rightleftharpoons}} \mathrm{AB}$$

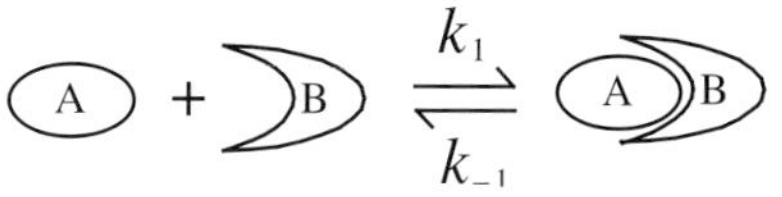

FIGURE 2.1. Molecular model of Equation 1.

$$K_B = \frac{k_1}{k_{-1}} \qquad \text{units: } \mathrm{M}^{-1} = \frac{\mathrm{M}^{-1}\mathrm{s}^{-1}}{\mathrm{s}^{-1}} \tag{3}$$

These relationships are useful when thinking about how affinity constants relate to kinetics. For example, an interaction with a low K_D (high affinity) will likely have a large k_1 and a small k_{-1} (i.e., A and B will associate rapidly and the AB complex will dissociate slowly). In contrast, an interaction with a high K_D (low affinity) will likely have a small k_1 and a large k_{-1} (i.e., A and B will associate slowly and the AB complex will dissociate rapidly). Importantly, the forward and reverse rate constants for an interaction do not need to be known to measure the affinity constant; as described below, affinity constants can be measured directly.

Relating affinity constants to concentrations of the interacting molecules. The following expressions relate affinity constants to the concentrations of the three components present in a bimolecular reaction at equilibrium (A, B, and AB):

$$K_D = \frac{[A][B]}{[AB]} \qquad \text{units: } \mathrm{M} = \frac{\mathrm{M}\bullet\mathrm{M}}{\mathrm{M}} \tag{4}$$

$$K_B = \frac{[AB]}{[A][B]} \qquad \text{units: } \mathrm{M}^{-1} = \frac{\mathrm{M}}{\mathrm{M}\bullet\mathrm{M}} \tag{5}$$

where [A], [B], and [AB] are the molar concentrations of these reaction components at equilibrium. K_D and K_B depend on the concentrations of all three species in the reaction. For simplicity, all of the remaining equations in this chapter will be written in terms of K_D; however, K_B can always be calculated because $K_B = 1/K_D$.

Transforming Equation 4 to simplify measuring $\mathbf{K_D}$. Although Equation 4 is central to the theory behind affinity constants, it is often not possible to accurately measure the concentrations of all three reaction components (A, B, and AB) while maintaining equilibrium. A more practical equation for measuring K_D can be derived from Equation 4 by rearranging terms and substituting for one of the variables. The key to this derivation is the concept that at equilibrium, the population of A molecules will be split between free A and A bound in AB complexes ($[A]_{Total} = [A] + [AB]$). The same is true of the B molecules ($[B]_{Total} = [B] + [AB]$). The derivation generates an equation that lacks a term for the concentration of free A (see Appendix 1 for derivation):

$$\boxed{\frac{[AB]}{[A]_{Total}} = \frac{[B]}{[B] + K_D}} \tag{6}$$

Here, $[AB]/[A]_{Total}$ is the fraction of A put into the reaction that is in the AB complex at equilibrium. This is commonly referred to as the *fraction bound.* [B] is the concentration of free B in the reaction at equilibrium. Measuring the K_D using Equation 6 involves keeping the concentration of A in the reaction constant and varying the amount of B added to reactions. At each point of B, the concentrations of both the AB complex and the free B are measured. The experimental and practical considerations for doing this are considered in the next section (Experimental Considerations).

Equation 6 defines a hyperbola of the form shown in Figure 2.2, where $[AB]/[A]_{Total}$ is plotted on the *Y* axis and [B] is plotted on the *X* axis. The fraction of A in the AB complex (fraction bound) increases from zero and asymptotically approaches 1.0 as the concentration of B in the reaction increases from zero and approaches infinity. The K_D is equal to the concentration of free B ([B]) at which half of the A added to the reaction is bound to B (i.e., $[AB]/[A]_{Total} = 0.5$), as indicated by the dashed lines in Figure 2.2. To best visualize the K_D, the *X* axis of the plot in Figure 2.2 is cut off at a value equal to 10 times the K_D. Recall that $K_B = 1/K_D$; therefore K_B can also be determined using Equation 6.

The theory behind determining a K_D is also illustrated by dividing the right half of Equation 6 by K_D, to obtain:

$$\frac{[AB]}{[A]_{Total}} = \frac{\frac{[B]}{K_D}}{\frac{[B]}{K_D} + 1} \qquad (7)$$

In this equation, when $[B] = K_D$, $[B]/K_D$ is equal to 1, and $[AB]/[A]_{Total}$ is equal to 0.5 (i.e., 1/[1+1]). When $[AB]/[A]_{Total}$ is plotted versus $[B]/K_D$, Equation 7 defines a hyperbolic curve where 50% of A is in the AB com-

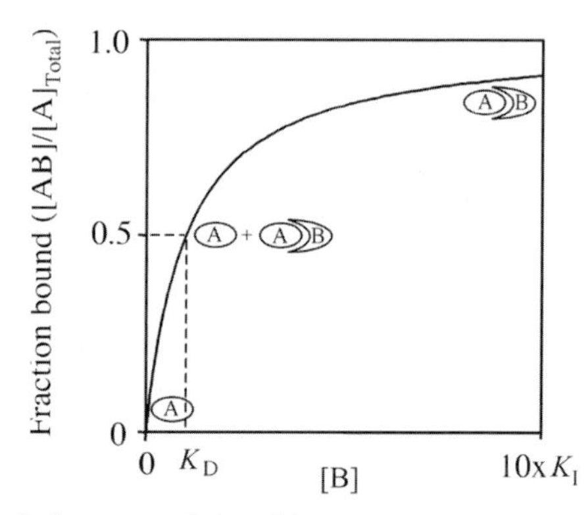

FIGURE 2.2. Hyperbolic curve defined by Equation 6. The dashed lines indicate that K_D is the concentration of B at which the fraction bound is 0.5.

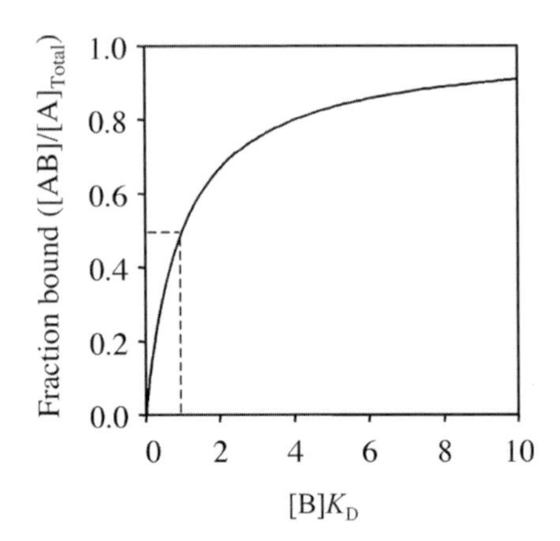

FIGURE 2.3. Bimolecular binding curve generated with Equation 7. The fraction bound equals 0.5 when $[B]/K_D$ equals 1.

plex when $[B]/K_D$ is equal to 1, as shown in Figure 2.3. This relationship is true for all bimolecular interactions irrespective of the actual value of the K_D.

Experimental Considerations

***Setting the concentrations of A and B: An experimental limitation to simplify measuring* K$_D$.** To experimentally measure K_D using Equation 6, a series of reactions are performed in which the concentration of A is kept constant, the concentration of B is varied, and the concentration of the AB complex (and possibly unbound B) is determined. B should be titrated such that the [B] at equilibrium spans a range that extends from well below the K_D to well above the K_D (minimally covering from tenfold below to tenfold above the K_D). Ironically, designing an experiment to accurately measure the K_D requires an initial estimate of the K_D for the interaction. If necessary, an estimate can be determined by performing a preliminary experiment in which B is titrated over a very large range. This initial titration of B may not allow an accurate K_D to be determined; however, it will provide an estimate of the K_D that can then be used in designing an experiment that will yield an optimal range of [B].

The total concentration of A is held constant while B is titrated. In principle, any concentration of A can be used; however, in practice it is simplest to determine a K_D when the concentration of A in the reactions is well below the K_D for the interaction ($[A]_{Total} << K_D$)—ideally, at least 100-fold below the K_D. Recall that the total concentration of B in a reaction is equal to the sum of the concentrations of free B and B bound in the AB complex ($[B]_{Total} = [B] + [AB]$). When $[A]_{Total} << K_D$ at all concentrations of B added to reactions, the amount of B in the AB complex is only a very small fraction of the total B. Therefore, free [B] is always much greater than [AB] and, hence, free [B] approximates $[B]_{Total}$ ($[B] \approx [B]_{Total}$). This simplifies performing experiments because, at equilibrium, free [B] does not have to be measured. Rather, the total concentration of B added to each reaction ($[B]_{Total}$) can be plotted on the X axis and the K_D can be determined using Equation 6. Alternatively, if $[A]_{Total}$ is not much less than K_D, free [B] must be measured at equilibrium in each reaction because under these conditions free [B] does not approximate $[B]_{Total}$.

Figure 2.4 shows four plots that demonstrate how the fraction bound ($[AB]/[A]_{Total}$) changes in relation to free [B] and $[B]_{Total}$ under two extreme conditions: $[A]_{Total} << K_D$ and $[A]_{Total} >> K_D$ (see also Simulation S2-1 on the Web site). Under both conditions, B is titrated at concentrations from well below to well above the K_D, which is 10 nM. All four plots in Figure 2.4 have $[AB]/[A]_{Total}$ on the Y axes, and either [B] or $[B]_{Total}$ on the X axes. The plots in Figure 2.4A were both generated under conditions where $[A]_{Total} << K_D$. The curves on the left and the right are identi-

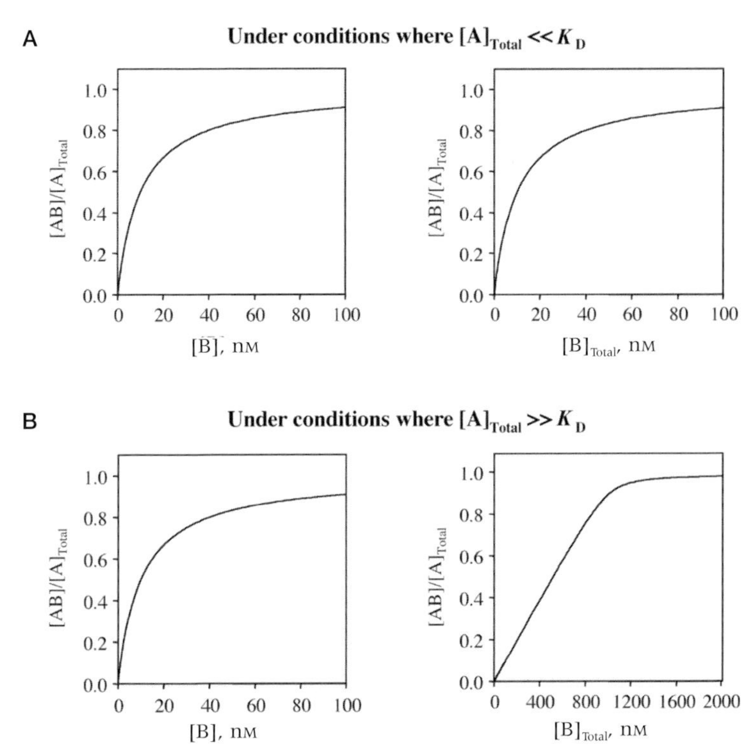

FIGURE 2.4. Bimolecular binding curves illustrating two experimental extremes: $[A]_{Total} << K_D$ and $[A]_{Total} >> K_D$. In all plots, the K_D is 10 nM. $[A]_{Total}$ is 0.1 nM in (*A*) and 1000 nM in (*B*).

cal because under these conditions, free [B] is equal to $[B]_{Total}$. In contrast, the plots in Figure 2.4B were generated under conditions where $[A]_{Total} >> K_D$, and therefore, free [B] does not equal $[B]_{Total}$. Here the curves on the left and the right are distinctly different. The curve on the left can be used to determine the K_D because free [B] is plotted on the X axis, which is required when $[A]_{Total}$ is not $< K_D$.

The plot on the right in Figure 2.4B has $[B]_{Total}$ plotted on the X axis. Although this plot cannot be used to determine a K_D, it is instructive to understand how $[B]_{Total}$ relates to the fraction bound when working under conditions where $[A]_{Total} >> K_D$. The curve does not appear hyperbolic, but instead increases linearly from a fraction bound of zero to a plateau value of 1.0. The linear region depicts reactions containing B at concentrations ($[B]_{Total}$) below $[A]_{Total}$, where nearly all of the B added to reactions is bound in AB complex and little is free at equilibrium. Under these conditions, $[AB]/[A]_{Total}$ increases nearly linearly with $[B]_{Total}$. The plateau region depicts reactions where $[B]_{Total}$ exceeds $[A]_{Total}$. Here, nearly all of the A is bound in AB complex and $[AB]/[A]_{Total}$ is close to 1.0. Note also that the range on the X axis in the right plot of Figure 2.4B is different from the other three plots; it takes a much higher $[B]_{Total}$ to saturate the A in the reaction and reach the plateau region when $[A]_{Total} >> K_D$.

In summary, the simplest way to measure a K_D is to set $[A]_{Total} << K_D$, plot $[B]_{Total}$ on the X axis, the $[AB]/[A]_{Total}$ on the Y axis, and fit the data with Equation 6. If these conditions cannot be met, the concentration of free [B] at equilibrium must be measured for each $[B]_{Total}$. For interactions in which the K_D is quite low (e.g., pM), accurately determining the K_D might be difficult because of issues with detecting the AB complex or measuring the concentrations of free [B]. In this case, $[A]_{Total}$ can be held as low as possible and an upper limit for the K_D can be measured. Alternatively, it might be possible to use competition assays to determine the K_D (as described in a later section of this chapter).

Developing an assay. An assay to accurately measure changes in the amount of AB complex is required and, if necessary, an assay to measure free [B]. In most assays, one biomolecule is labeled (e.g., radioactively or fluorescently) to facilitate detecting that molecule and the AB complex. It is important to emphasize that the accuracy of a measured KD depends on the accuracy with which the active concentrations of A and B are known (quantitating and assessing the activity and concentrations of biomolecules are discussed in Chapter 1). Similarly, it is important to accurately quantitate the AB complex, which often involves accounting for background in the measurement (also discussed in detail in Chapter 1). Finally, the K_D is affected by assay conditions such as pH, temperature, salt concentration, and the presence of divalent ions. These parameters should be considered when designing an experiment (also discussed in Chapter 1).

Establishing equilibrium. Experiments to measure affinity constants must be performed under equilibrium conditions. The simplest way to assess whether an interaction has reached equilibrium is to monitor the AB complex over time and determine when there is no net change in its concentration. This should be done with the lowest concentration of B that will be used in the experiment to measure K_D because the forward rate of binding will be slowest under this condition. From this time-course experiment, a single time point at which the reaction is at equilibrium can be chosen and used in reactions to measure the K_D.

Determining K_D when $[AB]/[A]_{Total}$ is measured. The following approach to determining a K_D follows directly from the theory and equations presented in the previous Theory and Equations section. At each concentration of B, the concentration of AB complex is quantitated and divided by the total concentration of A in the reaction to obtain $[AB]/[A]_{Total}$. Alternatively, if A is labeled, it may be possible to directly measure the fraction of A bound in the complex as opposed to measuring the concentration of AB and dividing by $[A]_{Total}$.

Data are plotted with the fraction of A bound in the complex ($[AB]/[A]_{Total}$) on the *Y* axis and the concentration of free B ([B]) on the *X* axis. Recall that when working under the conditions where $[A]_{Total} << K_D$, then $[B] \approx [B]_{Total}$ and the concentration of B added to reactions can be plotted on the *X* axis. To determine the K_D, the data points can be fit with a curve defined by Equation 6. Figure 2.5A shows a binding curve generated using Equation 6 (see also Simulation S2-2 on the Web site). The K_D

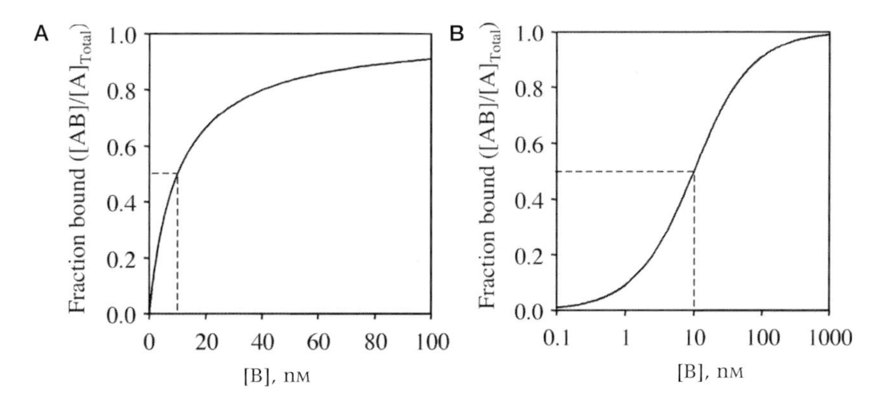

FIGURE 2.5. Bimolecular binding curves plotted on a linear *X* axis in (*A*) and a logarithmic *X* axis (*B*). The K_D is 10 nM in both plots.

is equal to the concentration of B at which the fraction bound is 0.5 (i.e., half of A is in the AB complex); 10 nM in this example. Computer programs can use nonlinear least-squares analysis to calculate the K_D given in Equation 6 (curve fitting is discussed in Chapter 7). Alternatively, draw a horizontal line from 0.5 on the *Y* axis over to the data and then a vertical line from this point down to the *X* axis. The point at which the vertical line crosses the *X* axis is the K_D (as illustrated in Fig. 2.5A).

The plot in Figure 2.5B shows a different way of visualizing the same binding curve. It is often easier to visualize the K_D when the data are plotted on a logarithmic *X* axis (this is called a semilogarithmic plot). In this case, the curve is sigmoidal, which allows the important region of the curve to be visualized when data are plotted over a broader range of B concentrations (note the different scales on the *X* axes of the linear plot in Fig. 2.5A and the semilogarithmic plot in Fig. 2.5B). Note also that it is not possible to plot a point lacking B (concentration of zero) on the semilogarithmic plot, because the log of zero equals negative infinity.

Illustration 2.1. Determining K_D for a protein–DNA interaction.

Sequence analysis reveals that the gene you are studying has a site quite similar, but not identical, to the consensus sequence bound by a known transcriptional activator protein. You want to determine the affinity with which this activator binds your newly identified site. In the following experimental description, the activator protein is designated P and the new DNA site is designated N.

$$N + P \rightleftharpoons NP$$

To begin, you purify P and determine its concentration and activity as described in Chapter 1. You also anneal and ^{32}P-label DNA oligonucleotides containing the binding site N. You decide to use electrophoretic mobility shift assays (EMSAs) to monitor the formation of the NP complex. It is known that P binds to its consensus site with a K_D of 20 nM, therefore, you estimate that P will bind to the new site N with a K_D no lower than 20 nM. You design a series of reactions in which the DNA concentration is held constant at 0.2 nM and P is titrated over a 500-fold range from 2 nM to 1000 nM. With this in mind, you perform a time-course experiment, which indicates that incubating N with 2 nM P for 1 hour is sufficient to reach equilibrium.

Satisfied that you have established conditions that will allow you to measure a K_D, you set up 20 reactions with different concentrations of P and incubate them for 1 hour. NP is separated from free N using native gel electrophoresis. The radioactivity in the bound and free bands is quantitated and the fraction bound is calculated for each reaction. You plot the fraction of the DNA bound in the NP complex versus the total concentration of P for each reaction (see plot below). Using a computer program and Equation 6, you determine the K_D to be 40 nM. You therefore conclude that the transcriptional activator protein P binds to site N in your gene with sufficient affinity to warrant further study.

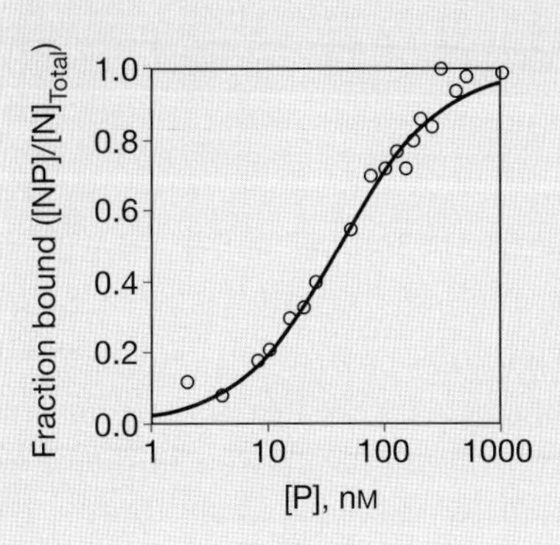

What to do when the fraction bound plateaus below 1. Equation 6 presumes that $[AB]/[A]_{Total}$ (fraction bound) approaches 1 as [B] is titrated well above the K_D. In many experiments, $[AB]/[A]_{Total}$ will actually approach a value less than 1; for example, if a portion of inactive A molecules is present in reactions. In these situations, an additional variable, f_{max}, is added to Equation 6, giving rise to (for derivation, see Appendix 1):

$$\frac{[AB]}{[A]_{Total}} = f_{max}\left(\frac{[B]}{[B] + K_D}\right) \tag{8}$$

Here, f_{max} represents the fraction bound at saturating [B]; it defines the plateau of the hyperbola. Hence, for experiments in which the data approach 1 as B is titrated up, f_{max} is equal to 1 and Equation 8 is identical to Equation 6. Most curve-fitting programs are capable of simultaneously determining f_{max} and K_D given Equation 8.

The plot in Figure 2.6 illustrates a binding curve generated with Equation 8. The curve plateaus at 0.7 (f_{max}); therefore, the K_D is equal to the concentration of B at which the fraction bound is 0.35 (half of the maximum binding). In general, it is preferable to use Equation 8 instead of Equation 6 for determining all K_D values. This will prevent inaccurate K_D measurements that arise from forcing curves to plateau at 1 when the data points do not.

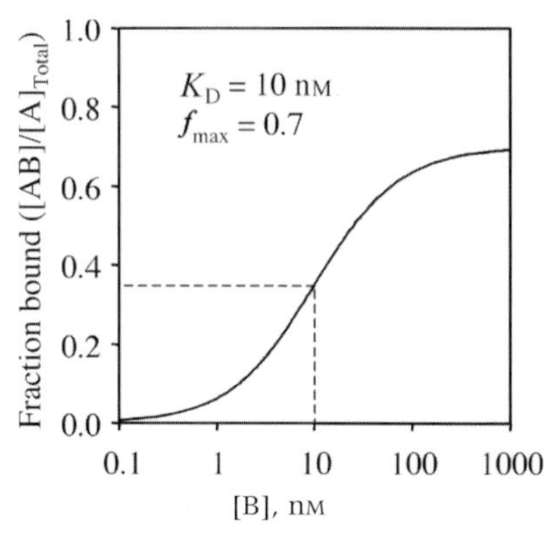

FIGURE 2.6. Bimolecular binding curve that plateaus at 0.7. The K_D (10 nM) is the concentration of B at which the fraction bound is half the maximum value (0.35).

Determining $\mathbf{K_D}$ ***by measuring AB.*** It is often most convenient to simply quantitate the amount of AB complex formed at different concentrations of B as opposed to determining the concentration of AB complex and dividing by $[A]_{Total}$. This approach is particularly useful when the labeled molecule is B (i.e., the molecule being titrated), but it can be used when A is labeled as well. When AB is measured as opposed to $[AB]/[A]_{Total}$, use the following equation to determine the K_D (for derivation, see Appendix 1):

$$AB = AB_{max}\left(\frac{[B]}{[B] + K_D}\right) \tag{9}$$

Here, AB is not a concentration. Instead, it is the amount of AB complex that is quantitated in a manner specific to the assay. For example, AB could be expressed in phosphorimager units, fluorescence emitted, or scintillation counts. AB_{max} is the maximum amount of AB complex that can form at saturating [B]; it will have the same units as AB. Unlike f_{max} and $[AB]/[A]_{Total}$ in Equation 8, AB_{max} and AB do not represent fractions of A that are in the AB complex. Rather, they are absolute expressions of the amount of the AB complex.

Consider the example shown in Figure 2.7, which can be applied to many systems. Here B was ^{32}P-labeled and the AB complex was quantitated by phosphorimagery. Phosphorimager units were plotted on the *Y* axis and [B] on the *X* axis. As the concentration of B increases, the phosphorimager units in the AB complex increase to a plateau defined by AB_{max}. Both the K_D and AB_{max} can be obtained by fitting the data with a nonlinear regression program using Equation 9. In this example, AB_{max} is 90,000 and the K_D is equal to 10 nM, which is equal to the concentration of B where the phosphorimager units are half the maximum value (45,000).

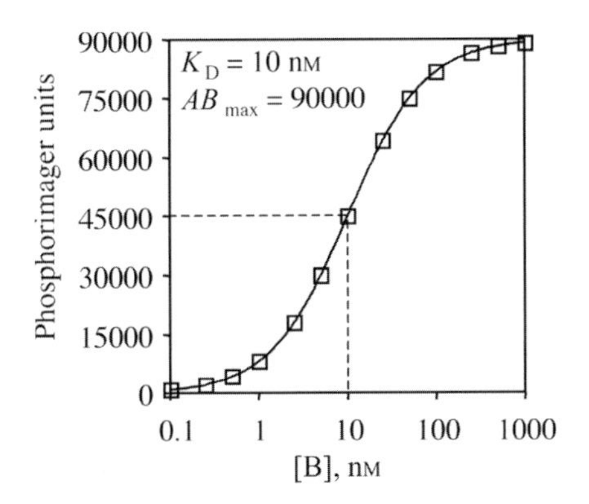

FIGURE 2.7. Bimolecular binding curve obtained from plotting phosphorimager units (i.e., the amount of AB) versus [B]. Shown are the K_D and AB_{max} used to generate the data points and curve.

Literature Example 2.1. Investigating the affinity of a protein–RNA interaction.
(Battle D.J. and Doudna J.A. 2001. RNA 7: 123–132.) Figures reprinted, with permission, from Battle and Dudna (2001).

This study examined the affinity of a protein for binding wild-type and mutant RNAs by measuring equilibrium dissociation constants. The protein studied was the stem-loop-binding protein (SLBP), which binds a unique stem-loop structure at the 3′ end of histone pre-mRNAs. The K_D for the interaction between SLBP and the RNA containing the most conserved elements of the histone mRNA stem-loop structure was determined. To assess which residues and regions in the RNA were important for high-affinity binding, K_D values were also measured using several RNAs with mutations and deletions, and the values were compared.

To assay protein–RNA complexes, the authors used nitrocellulose-filter-binding experiments in which the RNA was 5′ end-labeled with ^{32}P, and unlabeled SLBP protein was titrated. The concentration of the RNA in the reactions was kept below 50 pM, and the protein was titrated over an extremely broad range. (Note that the authors quantitatively determined the fractional activity of their SLBP preparations before performing experiments.) Reactions were incubated at 25°C for 1 hour to establish equilibrium. Binding data were plotted with the fraction bound on the *Y* axis and the concentration of SLBP on the *X* axis and fit with the equation:

$$f = (a - b)\left(\frac{[P]}{[P] + K_D}\right) + b$$

(Continued on following page)

Literature Example 2.1. *(Continued)*

where f is the fraction bound, $[P]$ is the active concentration of SLBP, and K_D is the equilibrium dissociation constant. The term a accounts for the fraction bound not approaching 1. The term b accounts for background binding of the RNA to the nitrocellulose filter; theoretically this should be zero. Therefore, this equation is similar to Equation 8. The data shown in the plot below are for the binding interaction between SLBP and the wild-type RNA. The K_D measured was 1.54 nM; the fraction bound approached 0.8.

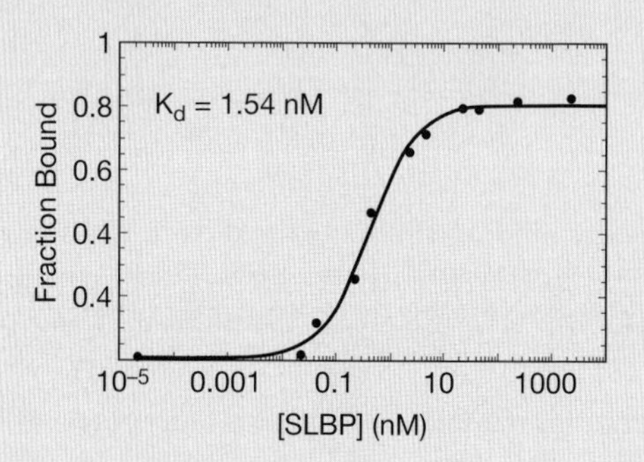

To investigate the determinants of the high affinity between SLBP and the stem-loop structure on the RNA, binding experiments were performed with RNAs containing mutations in the stem and the loop, as well as deletions of sequence surrounding the stem and loop. The plot below shows the binding data using five RNAs that contained mutations in the stem region. It is clear that some mutations enhanced the binding affinity (triangles) and some did not substantially affect the binding affinity (squares and diamonds), whereas others decreased the binding affinity (circles and inverted triangles).

Comparing the K_D values obtained with the wild-type RNA to those obtained with the mutant RNAs allowed the authors to determine which

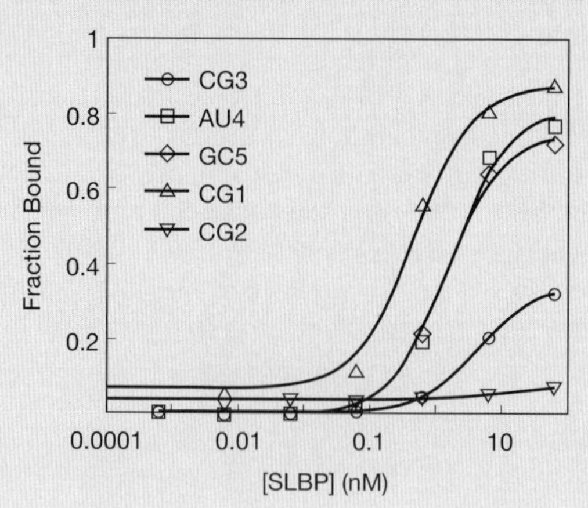

Summary: How to experimentally measure a K_D

1. Develop an assay to monitor AB.
2. Establish equilibrium.
3. Hold $[A]_{Total}$ constant (ideally, $[A]_{Total} \ll K_D$).
4. Titrate $[B]_{Total}$.
5. Measure either $[AB]/[A]_{Total}$ or the amount of AB.
6. If $[A]_{Total}$ is not $\ll K_D$, measure free [B].
7. Plot data and determine K_D (using either Equation 8 or 9).

USING COMPETITION TO ASSESS AFFINITY: IC_{50}

Competition assays can be used to quantitatively assess the relative affinities of two or more molecules for one target. For example, they can be used to determine how well mutant molecules bind to a target. In addition, competition assays can sometimes be used to determine the K_D for an extremely high-affinity interaction that is difficult to measure using the methods described in the preceding section.

Assessing affinity using competition assays requires having two or more molecules that bind to the same site on a target molecule. One of these molecules is often called a competitive inhibitor. Inhibitor does not necessarily refer to a biological or natural inhibitor. For example, a deletion mutant of a functional RNA can be tested for its ability to inhibit the binding of the wild-type RNA to a protein. In this case, the mutant RNA is used as a competitive inhibitor. In the following section, we describe how to use competition assays between two molecules to assess the affinities of each for a single target molecule.

Theory and Equations

Modeling competition. Competition between two molecules that bind the same site on a third molecule can be described by the expression (also see Fig. 2.8):

$$\mathrm{A + B + X} \begin{array}{l} \overset{K_{D(AB)}}{\rightleftharpoons} \mathrm{AB} \\ \underset{K_{D(AX)}}{\rightleftharpoons} \mathrm{AX} \end{array} \qquad (10)$$

where B and X each bind to the same site on A. Therefore, A can either bind B to form the AB complex or bind X to form the AX complex, but a single

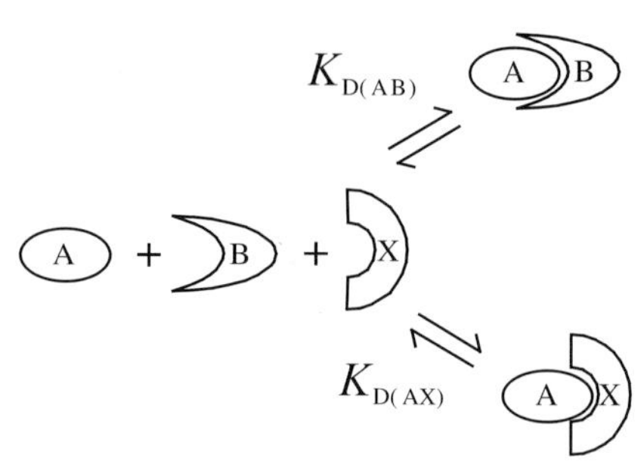

FIGURE 2.8. Molecular model of Equation 10.

molecule of A cannot simultaneously bind B and X. Moreover, B and X do not directly interact. We will consider X to be the competitor molecule. $K_{D(AB)}$ and $K_{D(AX)}$ are the equilibrium dissociation constants for the AB and AX complexes, respectively. The principles discussed below build on those introduced in the previous section. (Note that each equilibrium dissociation constant in Equation 10 relates to the equilibrium described by a pair of arrows.)

IC_{50} *definition and equations.* To ultimately determine the relative affinities with which B and X bind A, an IC_{50} is measured. The IC_{50} is equal to the concentration of X at which half of the maximal AB complex is present at equilibrium. In other words, the IC_{50} is equal to the concentration of the competitor molecule that reduces the binding of B to its target by 50%. The IC_{50} provides information about the affinity of a competitor molecule (X) for the target (A), compared to the affinity of the molecule it is competing with (B) for the target (A). Importantly, the IC_{50} is not the K_D for the interaction between the competitor molecule and its target.

Under conditions where the concentration of B is much greater than $K_{D(AB)}$ ($[B] >> K_{D(AB)}$) the following equation can be used to determine an IC_{50} (for derivation, see Appendix 1):

$$\boxed{\frac{[AB]}{[A]_{Total}} = 1 - \left(\frac{[X]}{[X] + IC_{50}}\right)} \tag{11}$$

Equation 11 defines a curve that starts at a fraction bound of 1 and asymptotically decreases to 0 as the concentration of X increases from 0 to infinity (see Fig. 2.9A). Therefore, in the absence of competitor molecule (X), all of the A in the reaction is bound by B in AB complexes. As the competitor molecule is titrated up, less AB complex forms and more AX complex is present. The IC_{50} is the concentration of X that reduces the fraction of AB complex from 1 to 0.5. This can be seen mathematically using Equation 11; when [X] is equal to IC_{50}, the right half of the equation is equal to 0.5.

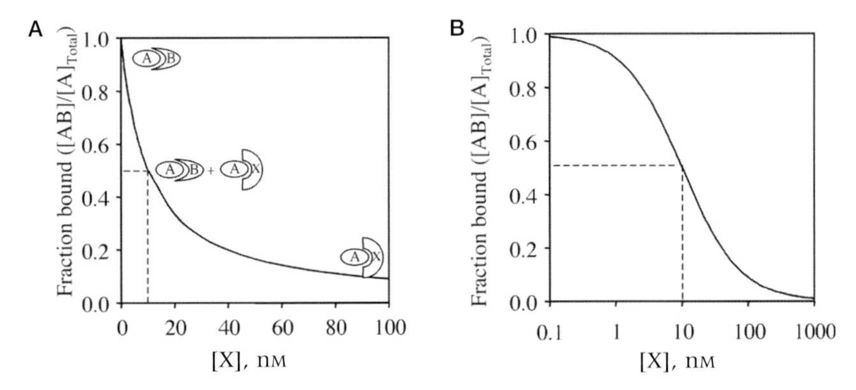

FIGURE 2.9. IC_{50} curves for competitive binding. *A* has a linear *X* axis and *B* has a logarithmic *X* axis. In both plots, the IC_{50} is 10 nM.

When the amount of AB complex is measured as opposed to $[AB]/[A]_{Total}$, the following equation can be used to determine the IC_{50}:

$$AB = AB_{max}\left(1 - \frac{[X]}{[X] + IC_{50}}\right) \tag{12}$$

Here, AB is not a concentration. Instead, it is the amount of AB complex that is quantitated in a manner specific to the assay. For example, AB could be expressed in phosphorimager units, fluorescence emitted, scintillation counts, etc. AB_{max} is the maximum amount of AB complex present before adding X; it will have the same units as AB.

Experimental Considerations

Many of the practical and experimental considerations previously discussed for measuring the K_D of a bimolecular interaction also apply to competition assays. To determine the IC_{50} using a competition experiment, the concentrations of A and B are held constant with $[B] >> K_{D(AB)}$, [A]. This condition allows the fraction of A in the AB complex ($[AB]/[A]_{Total}$) to be 1 in the absence of the competitor molecule X. X is titrated over a broad range from well below the concentration of B to well above. After the reaction reaches equilibrium, the fraction of A in the AB complex ($[AB]/[A]_{Total}$) is determined for each concentration of X. Typically, B is labeled (e.g., radioactively or fluorescently) and detected; however, A can be labeled as long as the method of detection can distinguish AB from AX.

The data are plotted with $[AB]/[A]_{Total}$ on the *Y* axis and the concentration of X on the *X* axis (note that $[X] \approx [X]_{Total}$ when $[X] >> [A]$). To determine the IC_{50}, the data points are fit with a curve defined by Equation 11. The plots in Figure 2.9 were generated using Equation 11 and an IC_{50} of

10 nM. To best visualize the IC_{50} in each plot, the X axes have different scales. A computer program can be used to fit the data to Equation 11 and solve for the IC_{50}. Alternatively, find 0.5 on the Y axis, draw a horizontal line over to the data, and then a vertical line down to the X axis. The point at which the line crosses the X axis is the IC_{50}.

Relating IC_{50} to Equilibrium Dissociation Constants

An IC_{50} is related to the equilibrium dissociation constants for the AX and AB complexes according to the following equation:

$$IC_{50} = K_{D(AX)} \left(1 + \frac{[B]}{K_{D(AB)}}\right) \tag{13}$$

When $[B] >> K_{D(AB)}$, Equation 13 simplifies to:

$$IC_{50} = \frac{K_{D(AX)}[B]}{K_{D(AB)}} \tag{14}$$

Rearranging Equation 14 gives an expression more useful for determining relative affinities between the two competing molecules (for derivation, see Appendix 1):

$$\boxed{\frac{K_{D(AX)}}{K_{D(AB)}} = \frac{IC_{50}}{[B]}} \tag{15}$$

Therefore, the relative affinities with which X and B bind A can be determined by simply dividing the measured IC_{50} by the concentration of B used in the experiment. For example, if the concentration of B is 10 nM and the IC_{50} measured is 100 nM, $K_{D(AX)}$ is tenfold greater than $K_{D(AB)}$. In other words, X binds A with tenfold lower affinity than B binds A.

Equation 15 can also be used to calculate either $K_{D(AX)}$ or $K_{D(AB)}$ from the IC_{50} when one of the two K_D values is known. This is particularly useful when one (but not both) of the K_D values is too low to measure directly. Consider an example in which $K_{D(AB)}$ is too low to directly measure; therefore, it will be determined using a competition assay with X. First, $K_{D(AX)}$ is measured directly using the equations and techniques described in the Bimolecular Interactions section of this chapter and is found to be 1 nM. Second, an IC_{50} is measured by titrating X into reactions containing 10 nM B and 0.1 nM A and is found to be 1 μM. From equation 15, $K_{D(AB)}$ is calculated to be 10 pM. If both $K_{D(AB)}$ and $K_{D(AX)}$ are too low to measure directly, the ratio of the two K_D values can still be determined from an IC_{50} experiment, but the K_D values themselves cannot be determined.

Literature Example 2.2. Using competition assays to assess the contributions of specific amino acids in the interaction between a cytokine and a cell surface.
(Clark K.D., Garczynski S.F., Arora A., Crim J.W., and Strand M.R. 2004. J. Biol. Chem. 279: 33246–33252.) Figure reprinted, with permission, from Clark et al. (2004).

Insects usually kill parasites by encapsulation, which entails attaching many layers of immune cells called hemocytes to the foreign target. Plasmatocyte-spreading peptide (PSP) is a 23-amino-acid cytokine that can activate hemocytes. In this study, the authors investigated the role of specific residues in PSP for binding to hemocytes. To do so, they created several PSPs with specific mutations and used competitive binding experiments and IC_{50} measurements to assess which amino acid residues in PSP are important for binding to hemocytes.

In their experiments, a native PSP peptide, $[^{125}I]$M12A, was held constant at a concentration of 100 pM (this can be considered the B molecule in Equation 10). Unlabeled mutant peptides (X molecules in Equation 10), including M12A as a control, were individually titrated from 1 pM to 10 μM to provide a range extending both below and above the concentration of $[^{125}I]$M12A. The binding target was 5 x 10^5 hemocytes (considered as the A molecules in Equation 10). The concentration of the target to which the peptides bind was not known with certainty but was well below the concentration of $[^{125}I]$M12A. Therefore, in the absence of competitor peptides, $[^{125}I]$M12A bound all sites on the hemocytes. Reactions were incubated for 3 hours at room temperature with regular vortexing to establish equilibrium. Cells were then washed thoroughly to remove any unbound peptides.

Binding was assessed by counting $[^{125}I]$M12A retained on the hemocytes compared to the total counts of $[^{125}I]$M12A added to the reactions. The fraction bound as a function of the concentration of the target could not be determined due to the use of whole cells instead of a purified protein. Rather, the data were plotted with the percentage of total $[^{125}I]$M12A bound on the *Y* axis versus the concentration of competitor peptide on the *X* axis for each of the competition experiments. Fitting the data with Equation 12 generated an IC_{50} value for each competitor peptide shown in the plot below.

Using this technique, the authors were able to determine that amino acids F3 and R13 were critical for binding to hemocytes because mutation of these amino acids dramatically increased the IC_{50}.

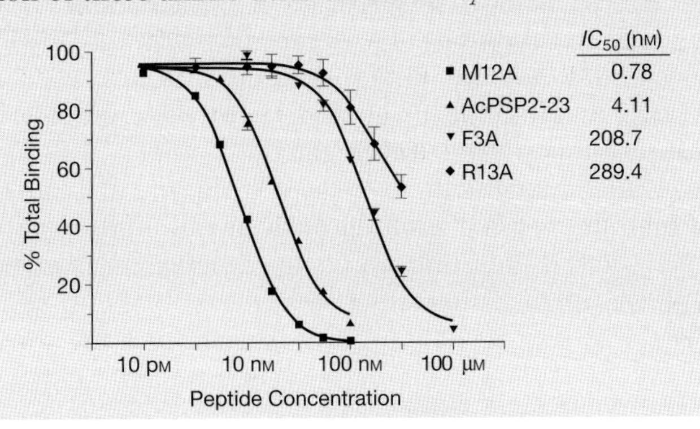

ASSESSING SPECIFICITY

K_D values are quantitative measurements of the *affinity*, or strength of an interaction, between two biomolecules. In contrast, *specificity* describes the relative affinities with which a biomolecule interacts with different partner molecules. For example, a protein that binds a DNA sequence with high affinity also exhibits high specificity only if it associates weakly with other DNA sequences. Alternatively, if that same protein bound many different DNA sequences with similar affinities, it would have a low specificity for binding DNA. Therefore, assessing specificity requires making multiple affinity measurements.

Consider again the bimolecular interaction A + B ↔ AB. A system for assessing the specificity of B for interacting with A versus other biomolecules related to A can be modeled by the following expressions:

$$\begin{aligned} A + B &\xrightleftharpoons{K_{D(AB)}} AB \\ A^* + B &\xrightleftharpoons{K_{D(A^*B)}} A^*B \end{aligned} \qquad (16)$$

where the lower expression depicts the binding reaction between B and any biomolecule that is related to A (represented by A*). For example, if A is a protein, A* could be a mutant protein, a protein from the same family, or a seemingly unrelated protein. Alternatively, if A is a DNA of a certain sequence, A* could be a DNA molecule of different sequence or a DNA oligonucleotide with a completely degenerate sequence. $K_{D(AB)}$ refers to the K_D for the interaction between B and A, and $K_{D(A^*B)}$ refers to the K_D for the interaction between B and any A* molecule.

To experimentally determine the specificity with which B binds to A, the K_D for the AB complex and $K_{D(A^*B)}$ values for one or more A*B complexes are determined. The $K_{D(A^*B)}$ values can be compared to $K_{D(AB)}$ to draw conclusions about the specificity of B for interacting with A versus A* molecules. If $K_{D(AB)}$ is much less than the $K_{D(A^*B)}$ values, B is specific in its interaction with A. In contrast, if $K_{D(AB)}$ does not differ greatly from any of the $K_{D(A^*B)}$ values, B does not bind specifically to A.

For any single A* molecule, dividing $K_{D(A^*B)}$ by $K_{D(AB)}$ provides a quantitative measure of specificity; the larger the number, the higher the specificity for B interacting with A versus A*. Importantly, knowing the specificity of B for binding A does not provide any indication of the specificity of A for binding B. A distinct set of experiments must be performed to assess the specificity of A interacting with B versus other B-related biomolecules (B*). It is also possible to assess specificity using competition assays rather than directly measuring $K_{D(A*B)}$ values. In Literature Example 2.2, competition was used to measure the affinity of cells for associating with a peptide (PSP) and a variety of mutant peptides. This approach is an example of assessing specificity by competition.

HOMODIMERIZATION (A + A ↔ AA)

Many biological molecules exist as oligomers, most commonly as dimers. Forming dimers from monomer units is governed by an affinity constant for dimerization. Whereas the equations in the first section of this chapter can be applied to heterodimer formation, the equations and analyses for homodimerization are unique.

Theory and Equations

Homodimerization reactions can be described by the following expression, where A represents monomers and AA represents dimers (also see Fig. 2.10):

$$A + A \rightleftharpoons AA \tag{17}$$

The equation relating the equilibrium dissociation constant to the concentrations of monomers and dimers is:

$$K_D = \frac{[A][A]}{[AA]} = \frac{[A]^2}{[AA]} \tag{18}$$

Importantly, there is a distinction between the concentration of free A monomers in the reaction and the total concentration of A added to the reaction. Consider the situation in which $[A]_{Total} >> K_D$. Most of the A will be in dimers (AA); however, the concentration of dimers can never be more than half of the concentration of total A added to the reaction (one mole of AA dimer contains two moles of A monomer). This concept is described by the following equation, where $[A]_{Total}$ is the total concentration of A added to the reaction:

$$[A]_{Total} = [A] + 2[AA] \tag{19}$$

$$A + A \rightleftharpoons AA$$

FIGURE 2.10. Molecular model of Equation 17.

The equation describing the K_D for homodimerization is derived by combining Equations 18 and 19 and using the quadratic equation to obtain (for derivation, see Appendix 1):

$$\frac{[AA]}{[A]_{Total}} = \frac{1}{2} + \frac{K_D - \sqrt{K_D^2 + 8K_D[A]_{Total}}}{8[A]_{Total}} \tag{20}$$

Although this equation looks intimidating, it can readily be applied to a data set to obtain the K_D for homodimerization. To do so, two parameters need to be known to calculate $[AA]/[A]_{Total}$ for each reaction: the concentration of A that went into the reaction ($[A]_{Total}$) and the concentration of the resulting dimers ([AA]).

Experimental Considerations

In measuring a K_D for homodimerization, many of the same experimental and practical considerations discussed earlier apply. First, an assay must be developed to separate monomers from dimers. A should be titrated in a series of reactions and $[AA]/[A]_{Total}$ determined molecule. The range of $[A]_{Total}$ tested should begin well below and end well above the K_D. Data are plotted with $[AA]/[A]_{Total}$ on the *Y* axis and $Total_{[A]}$ on the *X* axis. Equation 20 is then used to generate a curve that best fits the data and to determine the K_D using a nonlinear regression curve-fitting program. (Nonlinear regression is discussed in Chapter 7.) Figure 2.11 shows two curves generated from Equation 20 using a K_D of 10 nM. One is plotted with a linear

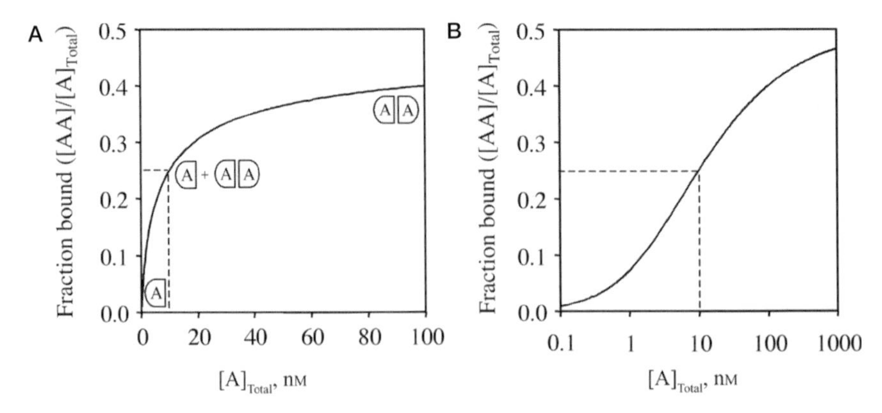

FIGURE 2.11. Binding curve for homodimers generated with Equation 20. *A* is plotted with a linear *X* axis and *B* is plotted with a logarithmic *X* axis. The K_D is 10 nM for each plot.

X axis (panel A) and the second with a logarithmic *X* axis (panel B) (see also Simulation S2-3 on the Web site).

Notice that as $[A]_{Total}$ increases, the curves approach 0.5. This is different from the plots describing the formation of AB complexes, which approach 1. This difference occurs because each dimer consists of two monomer subunits. Hence, the maximal concentration of dimers that can form will never exceed half of $[A]_{Total}$. The K_D for homodimerization is equal to the $[A]_{Total}$ at which $[AA]/[A]_{Total}$ is 0.25 (i.e., half of the maximum value). To estimate the K_D, find 0.25 on the *Y* axis, draw a horizontal line over to the data, and then a vertical line down from this point to the *X* axis. The point at which the line crosses the *X* axis is the K_D. The two *X* axes have different scales to best visualize the K_D.

Illustration 2.2. Determining the K_D for homodimerization of a protein.

You are studying a cytoplasmic protein named M. It is unknown whether M homodimerizes at the concentrations typically found in cells (~100 nM), although you have found that M can form homodimers in vitro. To address this question, you decide to measure the K_D for protein M homodimerization and compare it to the concentration of the protein in cells.

To assay for the formation of homodimers, you choose to use analytical gel-filtration chromatography. Because dimers are larger than monomers, they elute from the gel-filtration column earlier. You titrate purified protein M into 17 different reactions over a very broad concentration range (from 1 nM to 10 μM) because an approximation of the K_D for homodimerization is not known.

At concentrations of M well below the K_D, only monomers exist and you observe one peak coming off the column. As M is titrated up, you observe two peaks, one for monomers and one for dimers. As the concentration of M greatly exceeds the K_D, only dimers will exist and again you observe one peak. By integrating the volume under the dimer and monomer peaks, you are able to determine the fraction of protein in dimers for each concentration of M. You obtain the elution profile shown on page 42 on the left from a reaction containing a total of 80 nM of protein M. At this concentration, slightly more than half of the M added to the reaction is dimeric.

After running a column for each concentration of M and integrating under the dimer and monomer peaks, you plot the fraction of dimer formed versus the concentration of M added to reactions. Fitting the data with Equation 20 gives the curve shown in the plot on the right.

(Continued on following page)

Illustration 2.2. *(Continued)*

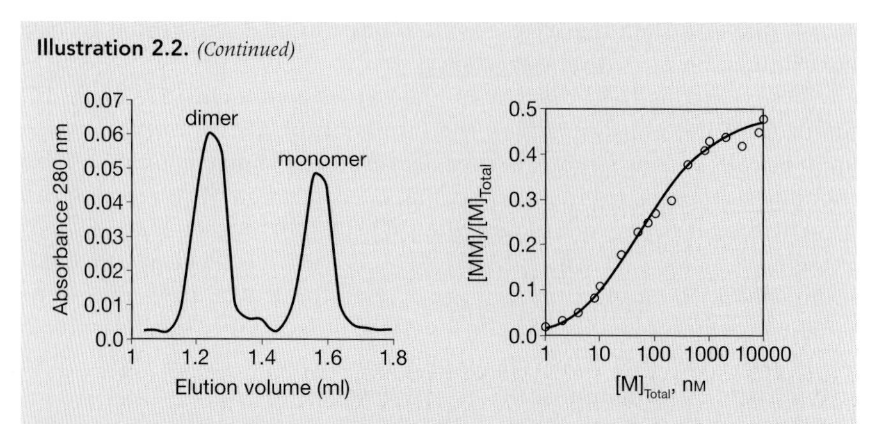

The K_D for homodimerization of protein M is 70 nM. You therefore conclude that, among the population of M in cells (~100 nM), a portion will likely exist as homodimers.

USING ENZYMATIC ASSAYS TO MEASURE AFFINITY CONSTANTS

If one of the biomolecules in a bimolecular binding reaction is an enzyme, it might be possible to use the enzymatic reaction to measure the binding affinity between the enzyme and a second biomolecule, perhaps a regulator. This is different from monitoring enzyme-substrate interactions, which are described in most biochemistry textbooks from the perspective of Michaelis-Menton kinetics, and we will not repeat this information here. Rather, we discuss situations in which the enzymatic reaction is used as a tool to measure an apparent K_D for the interaction between an enzyme and a regulatory molecule. A K_D measured in this way is referred to as an apparent K_D because the assay used to make the measurement is indirect (i.e., it does not directly monitor binding).

The following discussion relies on information presented in the Bimolecular Interactions section earlier in this chapter and relates most directly to the section on Determining K_D by Measuring AB and Equation 9. Here, we use E to stand for enzyme and R to stand for regulator (the molecule that binds the enzyme, but itself is not a substrate), which are analogous to A and B, respectively, in the context of the Bimolecular Interactions section.

Theory and Equations

Consider the following reaction in which an enzyme (E) binds a regulator (R) to form an ER complex that contains enzymatic activity.

$$\mathrm{E} + \mathrm{R} \underset{}{\overset{K_D}{\rightleftharpoons}} \mathrm{ER} \left(\begin{matrix} \mathrm{S} \\ \downarrow \\ \mathrm{P} \end{matrix} \right. \tag{21}$$

The enzyme in the ER complex can bind substrate (S) and catalyze a reaction to form product (P), and its enzymatic activity is dependent on the presence of (and its binding to) the regulatory molecule. The equation relating the equilibrium dissociation constant for the ER complex to the concentration of regulator, [R] is:

$$\mathrm{ER} = ER_{max} \left(\frac{[\mathrm{R}]}{[\mathrm{R}] + K_D} \right) \tag{22}$$

Here, ER is the amount of enzyme-regulator complex in a reaction at equilibrium and ER_{max} is the maximum amount of enzyme-regulator complex formed in the plateau region (i.e., at high R concentrations). Equation 22 is directly analogous to Equation 9.

When the enzymatic activity of E is used to monitor the binding reaction, ER in Equation 22 can be replaced with P (product made in a set time period) and ER_{max} can be replaced with P_{max} (the maximal product made in a set time period in the presence of saturating R) to obtain the following equation:

$$\mathrm{P} = P_{max} \left(\frac{[\mathrm{R}]}{[\mathrm{R}] + K_D} \right) \tag{23}$$

This equation is valid when ER, the active enzyme complex, produces product at a constant rate in the presence of a saturating amount of substrate.

For some enzymes, the regulator may not be required for enzymatic activity, but instead may increase or decrease the enzymatic activity. In other words, free enzyme can also convert substrate to product. In these cases, Equation 21 is modified and the reaction can be modeled as:

$$\left. \begin{matrix} \mathrm{S} \\ \downarrow \\ \mathrm{P} \end{matrix} \right) \mathrm{E} + \mathrm{R} \overset{K_D}{\rightleftharpoons} \mathrm{ER} \left(\begin{matrix} \mathrm{S} \\ \downarrow \\ \mathrm{P} \end{matrix} \right. \tag{24}$$

The equation relating the equilibrium dissociation constant for the ER complex to the concentration of R is (for derivation, see Appendix 1):

$$\boxed{\mathrm{E} + \mathrm{ER} = E_i + (ER_{max} - E_i) \left(\frac{[\mathrm{R}]}{[\mathrm{R}] + K_D} \right)} \tag{25}$$

Here, E and ER are the amounts of free enzyme and enzyme-regulator complex in a reaction at equilibrium, E_i is the amount of enzyme in the reaction before adding the regulator (i stands for initial), and ER_{max} is the maximum amount of enzyme-regulator complex formed in the plateau region (i.e., at high R concentrations).

Equation 25 can be rewritten in terms of the product produced by both E and the ER complex in the presence of saturating substrate, as follows:

$$P = P_i + (P_{max} - P_i)\left(\frac{[R]}{[R] + K_D}\right) \quad (26)$$

Here, P_i is the amount of product made in a set time period by the enzyme in the absence of the regulator and P_{max} is the amount of product made in a set time period when the enzyme is saturated with regulator. In cases where E alone (in the absence of R) does not have enzymatic activity, P_i is equal to 0, and Equation 26 reduces to Equation 23.

In Equation 26, the apparent K_D is equal to the concentration of R that gives rise to a P value that is halfway between P_i and P_{max}, which can be visualized on a plot of P versus [R]. Unlike other binding curves, if the regulator is an inhibitor of the enzyme, the curve will decrease as [R] increases. In this case, P_{max} will actually be the lowest value for the amount of product produced when the enzyme is saturated with R. This is illustrated in the following section.

Experimental Considerations

Measuring an apparent K_D using an enzymatic assay is similar in many ways to directly measuring binding. Most simply, a series of reactions is performed in which the concentration of E is kept constant and well below the K_D ($[E]_{Total} << K_D$) and the concentration of R is varied over a range that minimally extends from tenfold below the K_D to tenfold above the K_D. Instead of directly measuring binding, an enzymatic assay must be developed that will be capable of providing a relatively quick measurement of the amount of E + ER present in the reaction. To perform the assay, binding reactions containing E and R are allowed to come to equilibrium, and substrate is added at a concentration high enough to saturate the enzyme in the reaction. The enzymatic reaction is allowed to proceed for a short period of time such that the substrate concentration never decreases to a point at which it affects the amount of product made over the time course of the reaction. The amount of product produced by the enzyme in the absence of regulator should also be measured; this number will be used as a constant (P_i) when fitting data with Equation 26.

In designing the enzymatic assay, it is critical that the concentration of substrate be sufficiently high and the reaction time not too long, to ensure that the product increases linearly over time during the course of the reaction. To confirm these criteria, assemble a pilot reaction under conditions that give the greatest amount of enzymatic activity (either enzyme alone or in the presence of saturating R, depending on whether R decreases or increases the enzymatic activity). After adding a saturating concentration of substrate, stop portions of the pilot reaction at various time points, measure the amount of product, and plot it versus time. If the plot is linear through the time point planned for use in the experimental reactions, the concentration of substrate is sufficient and the reaction time is not too long.

Consider the example shown in Figure 2.12 in which increasing concentrations of an inhibitory molecule (R) were incubated with the enzyme (E) until reactions reached equilibrium. Saturating substrate was then added and the amount of product produced in 1 minute was quantitated and plotted on the Y axis and [R] on the X axis. In the absence of the inhibitor, the enzyme had substantial activity. As the concentration of R increased, the amount of product decreased to a plateau defined by P_{max}. Both the apparent K_D and P_{max} values were obtained by fitting the data with a nonlinear regression program using Equation 26 and the experimentally determined value for P_i (1000) as a constant in the equation. (Nonlinear regression is discussed in Chapter 7.) In this example, P_{max} is 100 and the apparent K_D is equal to 10 µM. The K_D is equal to the concentration of R at which the product made is halfway between that in the absence of R and that in the presence of saturating concentrations of R.

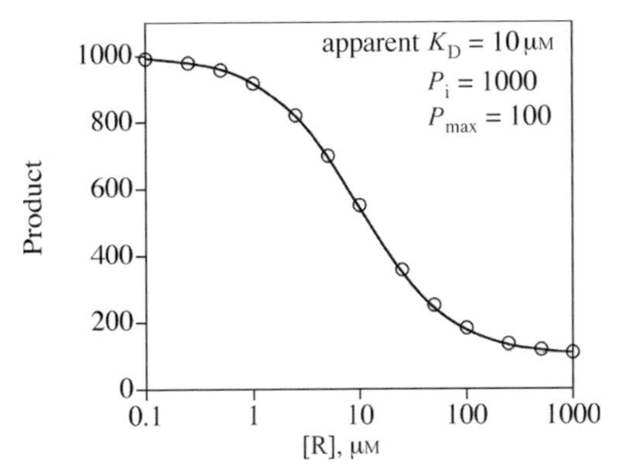

FIGURE 2.12. Bimolecular binding curve for the association of an inhibitor (R) with an enzyme. The product formed in a 1-minute reaction was plotted versus the concentration of R. Shown are the K_D, P_i, and P_{max} used to generate the data points and curve.

Literature Example 2.3. Using a kinase assay to measure the binding of calmodulin to smooth-muscle myosin light-chain kinase. *(Yuan T. and Vogel H.J. 1999. Protein Sci. 8: 113–121.) Figure reprinted, with permission, from Yuan and Vogel (1999).*

Smooth-muscle myosin light-chain kinase (MLCK) requires association with the calcium-bound calmodulin protein (CaM) for activity. In this study, the binding affinity between MLCK and CaM was measured by monitoring kinase activity in reactions containing increasing concentrations of calcium-bound CaM. Because MLCK is inactive in the absence of CaM, the reaction fits Equation 21 and can be diagrammed as:

$$\text{MLCK} + \text{CaM} \underset{}{\overset{K_D}{\rightleftharpoons}} \text{MLCK·CaM} \left(\begin{array}{l} \text{peptide} + [\gamma\text{-}^{32}\text{P}]\text{ATP} \\ \downarrow \\ [^{32}\text{P}]\text{peptide} + \text{ADP} \end{array} \right.$$

The authors also investigated the effect of replacing methionine residues in calmodulin with two unnatural amino acids (ethionine [Eth] and norleucine [Nle]) on the K_D of CaM for binding MLCK, as well as on the maximal level of kinase activity.

The experiments were performed by allowing MLCK-CaM complexes to form at multiple different concentrations of calcium-bound CaM. The reactions also contained saturating concentrations of a peptide that served as substrate for phosphorylation by MLCK. The kinase reaction was initiated with the addition of [γ-^{32}P]ATP. Reactions were allowed to proceed for 5 minutes and the amount of phosphorylated peptide was determined by acid precipitation and scintillation counting. Control experiments were performed to ensure that the reaction occurred at a constant rate over 5 minutes. The amount of [^{32}P]peptide produced at each concentration of CaM was normalized to the maximal amount of product made at saturating CaM to obtain a percentage. This percentage was plotted versus [CaM] and fit to the following equation to solve for K_D and [^{32}P]peptide$_{max}$:

$$[^{32}\text{P}]\text{peptide} = [^{32}\text{P}]\text{peptide}_{max} \left(\frac{[\text{CaM}]}{[\text{CaM}] + K_D} \right)$$

The plot below shows the data obtained for wild-type CaM (squares, dashed curve) and for two mutants, Nle-CaM (triangles, dashed-dotted curve) and Eth-CaM (circles, dotted curve). Note that the data are plotted on a logarithmic X axis. The apparent K_D for wild-type CaM binding MLCK is 25.4 nM, whereas that for both substituted CaM molecules is 39.6 nM. The authors concluded that the substitutions had only modest effects on the activity of the MLCK-CaM complex: The [^{32}P]peptide$_{max}$ values were 97% and 88% for Nle-CaM and Eth-CaM, respectively.

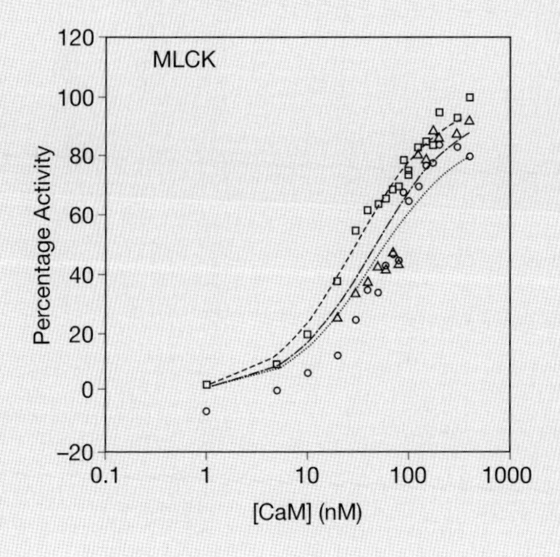

CHAPTER 3

Cooperativity in Binding

THIS CHAPTER DESCRIBES COOPERATIVITY and how to experimentally measure it for biological interactions. Cooperativity occurs in systems with three or more interacting partners. Quantitating cooperativity provides an understanding of how the binding of one molecule can affect the binding of others, either positively or negatively. Many examples of cooperative interactions exist in biology, including allosteric enzymes in which the binding of a cofactor influences the affinity with which substrate binds the enzyme, two proteins that bind DNA at adjacent or overlapping sites, ligands that bind receptors cooperatively, and the classic example discussed in most biochemistry text books of oxygen cooperatively binding to hemoglobin. There are also many examples of noncooperative interactions in biology in which each bimolecular binding reaction is independent and does not affect other interactions in the complex. For example, two proteins that bind neighboring sites on a DNA molecule but do not influence each other's binding are noncooperative. We present here hypothetical illustrations as well as examples from the literature to demonstrate practical aspects of assessing cooperativity and how it relates to affinity constants. Key equations are contained in boxes and derived in Appendix 1. In addition, computer simulations are provided on the Web site (http://kinetics.cshl.edu) that can be manipulated to help visualize concepts presented in the chapter.

Topics covered:

- Three-component interactions ($A + B + C \leftrightarrow ABC$)
- Positive, negative, and noncooperative binding
- The Hill plot and cooperativity ($A + nB \leftrightarrow AB_n$)

TERMS AND PRINCIPLES

Equilibrium is the state that is attained when all processes affecting a reaction are in balance. With respect to interactions, equilibrium exists when there is no overall change in the concentrations of the free and bound species in the reaction over time. Importantly, at equilibrium, a binding reaction is not static. Within the population, complexes are constantly forming while others are dissociating; however, there is no net change in bound versus free species.

Affinity describes the strength of an interaction between two molecules.

Dissociation constants (**K_D**) are quantitative parameters describing affinity that can be experimentally measured under equilibrium conditions. A K_D is expressed in units of M (moles/liter). K_D is inversely related to affinity; as affinity increases, K_D decreases.

Cooperativity occurs in systems with three or more interacting molecules when the binding of one molecule affects the binding of others, either positively or negatively. For example, in a three-component interaction ($A + B + C \leftrightarrow ABC$), cooperativity would exist if the binding of B to A affected the affinity of A for C. Cooperativity can be either positive or negative. Positive cooperativity exits when the presence of B increases the affinity of A for C (in this case, B can be thought of as an activator of the AC interaction). Negative cooperativity exits when the presence of B decreases the affinity of A for C (in this case, B can be thought of as a repressor of the AC interaction). Cooperativity in binding does not require that all of the interacting molecules be different. For example, two A molecules could cooperatively bind a molecule of B.

Cooperativity factor α describes the level of cooperativity that arises when one biomolecule affects the affinity with which two other biomolecules interact. Values of α are always greater than zero. When α is less than 1, a system has positive cooperativity. When α is greater than 1, a system has negative cooperativity. When α equals 1, a system lacks cooperativity. It is important to note that α is not the Hill coefficient, which is also used to describe cooperativity.

Hill plots are a means of assessing cooperativity in systems where a single molecule binds two or more of a second molecule ($A + nB \leftrightarrow AB_n$). Hill plots are most useful for determining whether a system has positive cooperativity.

THREE-COMPONENT INTERACTIONS ($A + B + C \leftrightarrow ABC$)

Many higher-order complexes (containing more than two components) are found in biology. Here, we consider the formation of complexes containing three different components. We discuss approaches to measuring the affinity constants governing trimolecular complexes, with a focus on assessing cooperativity.

Theory and Equations

Modeling trimolecular interactions. The model below describes the formation of ABC considering all possible bimolecular complexes that can form en route to ABC (AB, AC, and BC):

$$\begin{array}{ccc} \mathrm{A} + \mathrm{B} & \rightleftharpoons & \mathrm{AB} \\ + & \searrow\!\!\!\nwarrow & + \\ \mathrm{C} & \mathrm{BC} + \mathrm{A} & \mathrm{C} \\ \updownarrow & \searrow\!\!\!\nwarrow & \updownarrow \\ \mathrm{AC} + \mathrm{B} & \rightleftharpoons & \mathrm{ABC} \end{array} \qquad (1)$$

Figure 3.1 shows a model that pictorially represents the formation of ABC as described by Equation 1.

Contacts between all three molecules contribute to the formation and stability of the ABC complex. In many cases, however, one of the three bimolecular complexes forms only weakly at biological concentrations. For example, consider two proteins (B and C) that bind DNA (A) and also make protein–protein contacts with each other. Each protein alone binds to DNA with a K_D in the nanomolar range. Whereas the proteins can interact apart from DNA, their affinity for one another is quite low (K_D in the high micromolar range); therefore, at typical biological concentrations, they would not frequently be found in a stable complex off of DNA. In this situation, experiments to measure affinity constants and assess

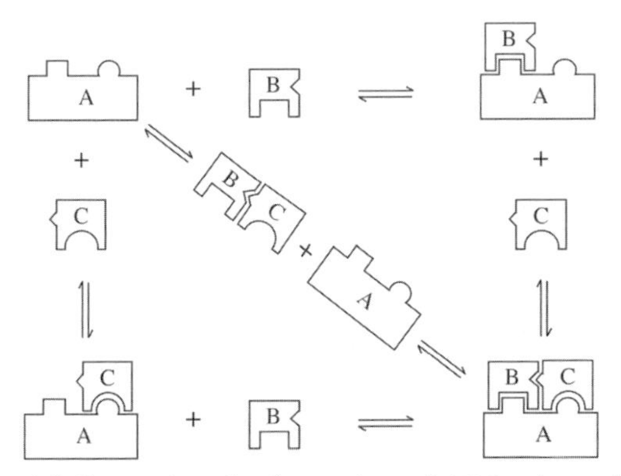

FIGURE 3.1. Model illustrating the formation of ABC where three bimolecular interactions (AB, AC, and BC) can form.

cooperativity can be performed with protein concentrations low enough to preclude the protein–protein interaction from occurring at any appreciable level off of DNA. Considering the protein–protein complex to be BC, these conditions allow a simpler version of Equation 1 to be written that considers the formation of only two of the bimolecular complexes (AB and AC):

$$\begin{array}{ccc} A + B & \rightleftharpoons & AB \\ + & & + \\ C & & C \\ \upharpoonleft\downharpoonright & & \upharpoonleft\downharpoonright \\ AC + B & \rightleftharpoons & ABC \end{array} \qquad (2)$$

A pictorial model showing the formation of ABC as described by Equation 2 is displayed in Figure 3.2.

It is far easier to understand and measure affinity constants and cooperativity for trimolecular interactions using the model in Equation 2, and many three-component interactions in biology can be modeled in this manner. In the following sections, we describe the theory, equations, and experimental considerations used in evaluating systems that fit the model shown in Equation 2.

Molecular mechanisms of cooperativity. Cooperativity occurs in systems with three interacting molecules when the binding of one biomolecule affects the binding of others, either positively or negatively. At the molecular level, there are two common mechanisms that could give rise to cooperativity in trimolecular interactions: (1) weak contacts between B

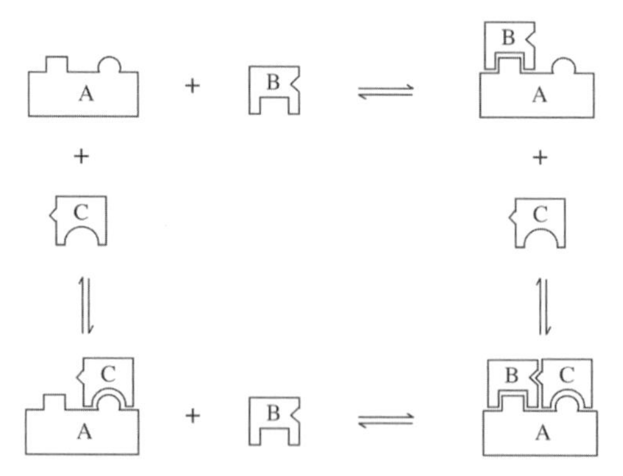

FIGURE 3.2. Model illustrating the formation of ABC where two bimolecular interactions (AB, AC) form. The BC complex does not form.

A

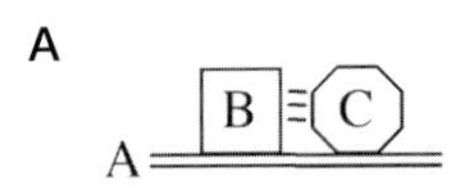

B

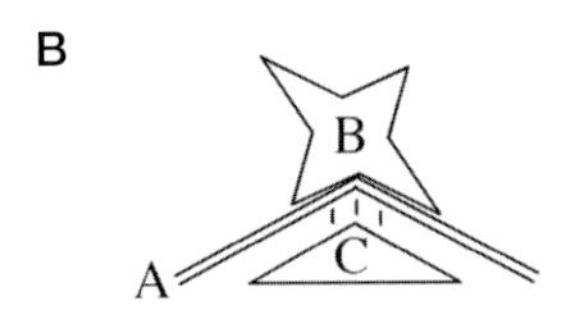

FIGURE 3.3. Models illustrating two possible mechanisms by which positive cooperativity occurs while forming an ABC complex. See text for a complete description.

and C that affect the overall formation of the ABC complex and (2) the binding of B (or C) to A, inducing a conformational change in A that alters its affinity for C (or B). Consider again the example of two proteins (B and C) binding with positive cooperativity to DNA (A). In Figure 3.3A, the horizontal lines between B and C represent protein–protein contacts that form when the proteins are bound to DNA. These contacts allow the ABC complex to assemble with positive cooperativity. Figure 3.3B also shows two proteins (B and C) that bind DNA (A) with positive cooperativity; however, they do through a different mechanism. In this example, B bends the DNA when it binds. This conformational change in the DNA then allows C to bind with higher affinity, leading to positive cooperativity. If the proteins interfered with one another while binding DNA or the conformational change caused by B decreased the affinity for C binding, then the cooperativity would be negative.

***The cooperativity factor* α.** Consider the following equations that describe the formation of ABC when both the AB and AC complexes can form, but BC does not:

$$
\begin{array}{ccc}
 & K_{D(AB)} & \\
A + B & \rightleftharpoons & AB \\
+ & & + \\
C & & C \\
K_{D(AC)} \ \upharpoonleft\downharpoonright & & \upharpoonleft\downharpoonright \ \alpha K_{D(AC)} \\
AC + B & \rightleftharpoons & ABC \\
 & \alpha K_{D(AB)} &
\end{array}
\tag{3}
$$

This model can also be depicted by the combination of two linear equations:

$$\begin{array}{c} A + B \underset{}{\overset{K_{D(AB)}}{\rightleftharpoons}} AB + C \underset{}{\overset{\alpha K_{D(AC)}}{\rightleftharpoons}} ABC \\ \\ A + C \underset{}{\overset{K_{D(AC)}}{\rightleftharpoons}} AC + B \underset{}{\overset{\alpha K_{D(AB)}}{\rightleftharpoons}} ABC \end{array} \tag{4}$$

$K_{D(AB)}$ is the equilibrium dissociation constant for B binding A and $K_{D(AC)}$ is the equilibrium dissociation constant for C binding A.

The term α, which has not appeared in the binding equations presented in this volume thus far, describes the cooperativity in the system. For example, the effect of B on the K_D for the interaction of A with C is represented by α, in that $K_{D(AC)}$ becomes $\alpha K_{D(AC)}$ in the presence of B. If there is positive cooperativity, α is less than 1; the presence of B increases the affinity of A for C (e.g., $\alpha K_{D(AC)} < K_{D(AC)}$). If there is negative cooperativity, α is greater than 1; the presence of B decreases the affinity of A for C (e.g., $\alpha K_{D(AC)} > K_{D(AC)}$). If there is no cooperativity in the system, α is equal to 1 and $\alpha K_{D(AC)}$ is equal to $K_{D(AC)}$. These concepts are summarized in Table 3.1.

It is important to understand that although present twice in Equations 3 and 4, α is a single constant with one value for a given system. In other words, if B affects the affinity of A for C by a factor of α, then C affects the affinity of A for B by the same factor of α (i.e., $\alpha K_{D(AC)}$ and $\alpha K_{D(AB)}$ can have different values, but α must be the same). To illustrate this point, consider the path to ABC outlined by the top and right sides of the model in Equation 3. The equilibrium dissociation constant describing this pathway equals $K_{D(AB)}\alpha K_{D(AC)}$ (the product of the K_D values for the two individual reactions). Similarly, the equilibrium dissociation constant describing the path to ABC outlined by the left and bottom sides of the model equals $K_{D(AC)}\alpha K_{D(AB)}$. The overall equilibrium dissociation constant describing the formation of ABC is independent of the order in which components assemble; therefore, $K_{D(AB)}\alpha K_{D(AC)} = K_{D(AC)}\alpha K_{D(AB)}$. For this expression to be true, the α values on each side of the equation must be equal.

TABLE 3.1. How cooperativity relates to binding affinity

Value for α	Cooperativity	Effect on affinity	Effect on K_D
$\alpha < 1$	positive	increase	decrease
$\alpha = 1$	none	none	none
$\alpha > 1$	negative	decrease	increase

Determining α. The value of α for a three-component interaction can be determined by measuring the K_D values for one of the bimolecular interactions in both the absence and presence of the third molecule. In the examples below, C is the third molecule, which can be considered a regulator of the interaction between A and B. To measure α for the formation of ABC, two equilibrium dissociation constants must be measured: the K_D for the binding of B to A in the absence of C ($K_{D(AB)}$) and the K_D for the binding of B to A in the presence of saturating C ($\alpha K_{D(AB)}$). The expressions for these two bimolecular interactions are shown below:

$$A + B \underset{}{\overset{K_{D(AB)}}{\rightleftharpoons}} AB \tag{5}$$

$$AC + B \underset{}{\overset{\alpha K_{D(AB)}}{\rightleftharpoons}} ABC \tag{6}$$

Remember that $K_{D(AC)}$ and $\alpha K_{D(AC)}$ could also be measured and the same value for α and conclusions regarding cooperativity should be obtained.

The theory and equations presented in Chapter 2 for measuring K_D values in bimolecular systems apply to measuring $K_{D(AB)}$ and $\alpha K_{D(AB)}$ in trimolecular systems as well. It may be helpful to refer back to Chapter 2 for a more detailed description of the theory behind the following equations.

The equation for determining $K_{D(AB)}$ relates the concentrations of the reaction components to $K_{D(AB)}$ in the following manner (for derviation, see Appendix 1):

$$\boxed{\frac{[AB]}{[A]_{Total}} = \frac{[B]}{[B]+K_{D(AB)}}} \tag{7}$$

To measure $K_{D(AB)}$, B is titrated into a series of reactions and $[AB]/[A]_{Total}$ (and, if necessary, free [B]) is determined at equilibrium. When $[AB]/[A]_{Total}$ is plotted versus [B], Equation 7 defines a hyperbolic curve that approaches 1. $K_{D(AB)}$ is the concentration of B at which $[AB]/[A]_{Total}$ is 0.5.

Measuring $\alpha K_{D(AB)}$ is similar, except that C is present at concentrations high enough to saturate all the A in each reaction. This situation mimics a bimolecular system in which the AC complex acts as a single entity as B is titrated into reactions to form ABC (as shown in Equation 6). $\alpha K_{D(AB)}$ is the equilibrium dissociation constant for this interaction and can be determined using the following equation (see Appendix 1 for derivation):

$$\boxed{\frac{[ABC]}{[A]_{Total}} = \frac{[B]}{[B] + \alpha K_{D(AB)}}} \tag{8}$$

When $[ABC]/[A]_{Total}$ is plotted versus [B], $\alpha K_{D(AB)}$ is the concentration of B at which $[ABC]/[A]_{Total}$ is 0.5.

Once $K_{D(AB)}$ and $\alpha K_{D(AB)}$ (or $K_{D(AC)}$ and $\alpha K_{D(AC)}$) are measured, α can be calculated.

$$\alpha = \frac{\alpha K_{D(AB)}}{K_{D(AB)}} \tag{9}$$

Experimental Considerations

Measuring affinity and assessing cooperativity in trimolecular systems involves measuring the K_D values for two separate bimolecular interactions using Equations 7 and 8. Recall that in Chapter 2, we discussed in detail the experimental approaches to measuring K_D values for bimolecular interactions. The box below contains a summary of the key points; however, it may be helpful to refer back to Chapter 2 for further explanation.

Measuring a K_D (summarized from Chapter 2)

1. Develop an assay to monitor AB.
2. Establish equilibrium.
3. Hold $[A]_{Total}$ constant (ideally, $[A]_{Total} << K_D$).
4. Titrate $[B]_{Total}$.
5. Measure either $[AB]/[A]_{Total}$ or the amount of AB.
6. If $[A]_{Total}$ is not $<< K_D$, measure free [B].
7. Plot data and determine K_D.

Assay development. Typically, when measuring K_D values in trimolecular systems and determining α, either molecule A or B is labeled. If A is labeled, an assay must be developed that can distinguish (1) AB from free A and (2) ABC from AC. If B is labeled, an assay must be developed that can distinguish (1) AB from free B and (2) ABC from free B.

Determining α when $[AB]/[A]_{Total}$ and $[ABC]/[A]_{Total}$ are measured. To measure $K_{D(AB)}$, the concentration of A is held constant and well below the K_D, and the concentration of B is titrated over a broad range. $[AB]/[A]_{Total}$ is measured and $K_{D(AB)}$ is determined by plotting $[AB]/[A]_{Total}$ versus [B] and fitting the data to Equation 7. Fitting these types of curves using nonlinear regression is discussed in Chapter 7.

$\alpha K_{D(AB)}$ is measured using a similar approach, with one key exception. Molecule C is included in reactions at a concentration that allows it to saturate all of the A in the reaction. The presence of unbound C in reactions is not a problem as long as its concentration is not high enough to form the BC complex (this could be confirmed experimentally). $[ABC]/[A]_{Total}$ is measured for each reaction as the concentration of B is titrated, and $\alpha K_{D(AB)}$ is determined by plotting $[ABC]/[A]_{Total}$ versus [B] and fitting the data to Equation 8.

The cooperativity factor α is calculated from $K_{D(AB)}$ and $\alpha K_{D(AB)}$ using Equation 9. If $\alpha K_{D(AB)}$ (measured in the presence of C) is equal to $K_{D(AB)}$ (measured in the absence of C), then α is equal to 1 and the formation of the trimolecular complex is noncooperative; the presence of C does not affect the affinity with which A and B interact. If $\alpha K_{D(AB)}$ is less than $K_{D(AB)}$, then α is less than 1 and there is positive cooperativity; C increases the affinity between A and B. Conversely, if $\alpha K_{D(AB)}$ is greater than $K_{D(AB)}$, then α is greater than 1 and there is negative cooperativity; C decreases the affinity between A and B. Importantly, $K_{D(AC)}$ and $\alpha K_{D(AC)}$ could also be measured to assess cooperativity. Although their values will likely be different from those of $K_{D(AB)}$ and $\alpha K_{D(AB)}$, their ratio and therefore α should be the same. If the values obtained for α are different, however, the simple model in Equation 3 does not adequately represent the reaction, and it likely includes additional steps or involves higher-order complexes.

The curves in Figure 3.4 were generated using Equations 7 and 8 and together illustrate positive cooperativity (see also Simulation S3-1). The right curve is for $K_{D(AB)}$ and shows the affinity between A and B in the absence of C. The fraction of A in the AB complex ($[AB]/[A]_{Total}$) is plotted on the right *Y* axis. The left curve is for $\alpha K_{D(AB)}$ and shows the affinity

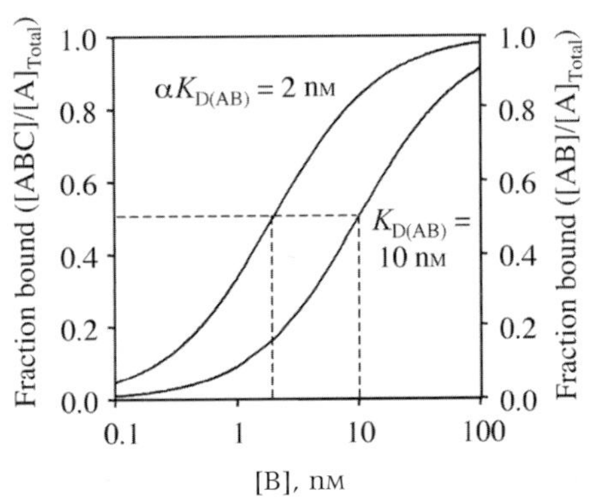

FIGURE 3.4. Binding curves measuring $K_{D(AB)}$ (10 nM) and $\alpha K_{D(AB)}$ (2 nM). Plotted with a logarithmic *X* axis.

between A and B in the presence of C. The fraction of A in the ABC complex ($[ABC]/[A]_{Total}$) is plotted on the left *Y* axis. The concentration of B is plotted on a logarithmic *X* axis. As seen in the plot, $K_{D(AB)}$ is equal to 10 nM and $\alpha K_{D(AB)}$ is equal to 2 nM. Here, C increased the affinity of B for A five-fold because, in the presence of C, the K_D for the interaction dropped from 10 nM to 2 nM. The value of α is 0.2, reflecting positive cooperativity. If the formation of ABC were noncooperative, then α would equal 1 and the curve for $\alpha K_{D(AB)}$ would overlay that for $K_{D(AB)}$. If the formation of ABC were negatively cooperative, then α would be greater than 1 and the curve for $\alpha K_{D(AB)}$ would shift to the right with respect to that for $K_{D(AB)}$.

Determining* α *by measuring AB and ABC. Often, it is most convenient to quantitate the amount of AB or ABC complex formed at different concentrations of B as opposed to measuring the concentration of AB or ABC complex ([AB] or [ABC]) and dividing by $[A]_{Total}$. This approach is particularly useful when the labeled molecule is B (i.e., the molecule being titrated), but it can be used when A is labeled as well. When the amounts of AB and ABC are measured, the following equations should be used to determine $K_{D(AB)}$ and $\alpha K_{D(AB)}$ (for derivations, see Appendix 1):

$$\mathrm{AB} = AB_{\max}\left(\frac{[\mathrm{B}]}{[\mathrm{B}]+K_{\mathrm{D(AB)}}}\right) \tag{10}$$

$$\mathrm{ABC} = ABC_{\max}\left(\frac{[\mathrm{B}]}{[\mathrm{B}]+\alpha K_{\mathrm{D(AB)}}}\right) \tag{11}$$

Here, AB and ABC are not concentrations, but amounts quantitated in a manner specific to the assay. For example, AB and ABC could be expressed in phosphorimager units, fluorescence emitted, scintillation counts, etc. AB_{max} and ABC_{max} are the maximum amount of AB and ABC complexes that can form, and they will have the same units as AB and ABC. An example of plotting and fitting this type of data is shown in Figure 2.7 in Chapter 2. Once $K_{D(AB)}$ and $\alpha K_{D(AB)}$ are measured, the cooperativity factor α is calculated using Equation 9.

Summary: How to experimentally measure α

1a. Hold $[A]_{Total}$ constant (ideally, $[A]_{Total} << K_{D(AB)}$).
b. Titrate $[B]_{Total}$.
c. Measure $[AB]/[A]_{Total}$ or the amount of AB.

d. If $[A]_{Total}$ is not << K_D, measure free [B].
e. Plot data and determine $K_{D(AB)}$ (using Equation 7 or 10).
2a. Hold $[A]_{Total}$ constant (ideally, $[A]_{Total} << \alpha K_{D(AB)}$).
b. Hold $[C]_{Total}$ constant (such that it saturates A).
c. Titrate $[B]_{Total}$.
d. Measure $[ABC]/[A]_{Total}$ or the amount of ABC.
e. If $[A]_{Total}$ is not << K_D, measure free [B].
f. Plot data and determine $\alpha K_{D(AB)}$ (using Equation 8 or 11).
3. Calculate α using Equation 9.

Illustration 3.1. Measuring cooperativity in a trimolecular complex containing two proteins and an RNA.

You are studying two proteins (F and G) that bind unique but adjacent sites on an RNA molecule (R) in order to stabilize its structure. The proteins make contact with each other once bound to the RNA, but you have not detected stable protein–protein interactions between F and G in the absence of the RNA. You want to know whether F and G bind the RNA cooperatively.

The expressions for the formation of the RFG complex (similar to Equation 4) can be written as:

$$R + F \underset{}{\overset{K_{D(RF)}}{\rightleftharpoons}} RF + G \overset{\alpha K_{D(RG)}}{\rightleftharpoons} RFG$$

$$R + G \overset{K_{D(RG)}}{\rightleftharpoons} RG + F \overset{\alpha K_{D(RF)}}{\rightleftharpoons} RFG$$

To determine if there is cooperativity, you need to measure two of the four equilibrium dissociation constants shown: either $K_{D(RF)}$ and $\alpha K_{D(RF)}$ or $K_{D(RG)}$ and $\alpha K_{D(RG)}$. You choose to measure $K_{D(RG)}$ and $\alpha K_{D(RG)}$.

The protein–RNA complexes can be detected using RNase T1 protection. RNase T1 cleaves single-stranded RNA 3′ of guanosine residues. When proteins F and G bind to the RNA, specific bands are protected from RNase T1 digestion. To determine the fraction of RNA bound in a protein–RNA complex, the level of protection can be quantitated. This technique enables the RG, RF, and RFG complexes to be detected.

(Continued on following page)

Illustration 3.1. *(Continued)*

You begin by ^{32}P-labeling the RNA at the 5′ end and, in a series of binding reactions, hold its concentration constant at 10 pM (presumed to be well below the K_D values for the RF and RG complexes). To first measure $K_{D(RG)}$, G is titrated from 0.1 nM to 1 μM into the reactions, which are then incubated at 30°C to allow binding to R and to establish equilibrium. The fraction of R in the RG complex ($[RG]/[R]_{Total}$) is determined at every concentration of G. You then plot these values versus the concentration of G (circles) and fit the data with the following equation:

$$\frac{[RG]}{[R]_{Total}} = \frac{[G]}{[G]+K_{D(RG)}}$$

giving rise to the curve on the left in the plot below. $K_{D(RG)}$ is 8 nM.

Next, to determine $\alpha K_{D(RG)}$, you perform a similar experiment, again holding R constant at 10 pM, but now adding F to the reactions at a concentration of 5 μM, which is sufficient to saturate the RNA. Note that excess F in the reaction is not a concern, because F does not interact with G off of RNA at concentrations used in this experiment. Then, titrate G into reactions over the same concentration range as in the first experiment (without G). After determining the fraction of R in the RFG complex for every concentration of G, you plot these values versus the concentration of G (triangles) and fit the data with the following equation:

$$\frac{[RFG]}{[R]_{Total}} = \frac{[G]}{[G]+\alpha K_{D(RG)}}$$

This results in the curve on the right where $\alpha K_{D(RG)}$ is 60 nM.

Clearly, the presence of F decreases the affinity of G for the RNA ($\alpha K_{D(RG)}$ is greater than $K_{D(RG)}$). You calculate the value of α to be 7.5. This number is greater than 1, allowing you to conclude that proteins F and G bind the RNA with negative cooperativity.

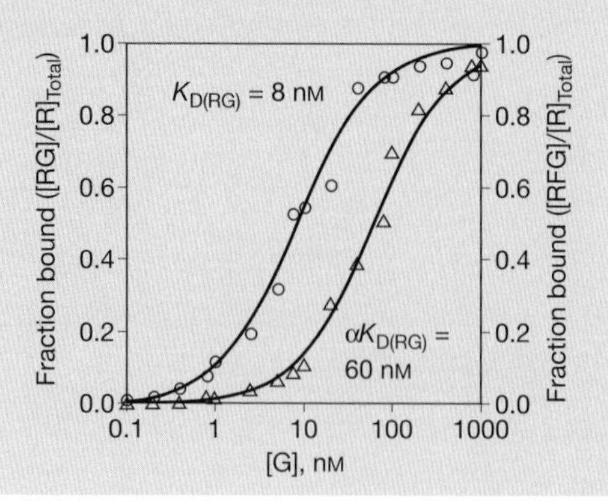

THE HILL PLOT AND COOPERATIVITY ($A + nB \leftrightarrow AB_n$)

In some biological systems, two or more identical molecules bind to distinct sites on a single target molecule, for example, a receptor with multiple binding sites for the same ligand. Hill plots are commonly used to assess cooperativity in these situations. Here, we describe various approaches to using Hill plots and address the limitations of their use.

Theory and Equations

Modeling $A + 2B \leftrightarrow AB_2$. The following expression considers the situation in which two molecules of one type bind to a third:

$$\begin{array}{ccc} A + B & \rightleftharpoons & AB \\ + & & + \\ B & & B \\ \upharpoonleft\downharpoonright & & \upharpoonleft\downharpoonright \\ AB + B & \rightleftharpoons & AB_2 \end{array} \tag{12}$$

This can also be described by the linear equation:

$$A + B \rightleftharpoons AB + B \rightleftharpoons AB_2 \tag{13}$$

The analysis and equations used to assess cooperativity in forming AB_2 differ from those used in studying trimolecular complexes. This is because most experimental techniques cannot distinguish between B interacting with site 1 or with site 2 on A. Moreover, it is typically not possible to saturate A with B at only one site and then subsequently measure the binding of B to the other site. Therefore, different equations must be used to determine whether there is cooperativity in forming AB_2. In other words, does the binding of the first molecule of B to A affect the affinity with which the second molecule of B binds?

The Hill equation. The Hill equation and Hill plots are commonly used to determine whether there is cooperativity in systems containing multiple B molecules binding to A. The Hill equation provides a useful means of assessing cooperativity as long as its limitations are understood. For example, the Hill equation can provide clear evidence for positive cooperativity; however, it cannot be used to definitively show negative cooperativity or noncooperative binding without also knowing the relative affinities of the different sites on A for binding B. In addition, the Hill equation cannot provide accurate measurements of the K_D values for the interaction of B at any site on A, the number of B molecules that bind to A, or the level of cooperativity described by α. Given these limitations, the Hill equation is most useful for determining whether there is positive cooperativity in systems where two or more molecules of one type bind to a single target.

The limitations of the Hill equation result from the primary assumption made during its derivation, namely, that perfect cooperativity exists in a system. From an experimental perspective, the problem with this assumption is that perfect cooperativity does not occur in biology. However, if perfect cooperativity for n molecules of B binding to A is assumed, then the following expression can be written:

$$A + nB \rightleftharpoons AB_n \tag{14}$$

Given perfect cooperativity, no AB complexes with fewer than n molecules of B ever exist, and therefore, these terms are not present in Equation 14.

The relationship between the equilibrium dissociation constant and concentrations of molecules in Equation 14 is:

$$K_D = \frac{[A][B]^n}{[AB_n]} \tag{15}$$

The Hill equation can be derived from Equation 15 (for derivation, see Appendix 1):

$$\boxed{\log\left(\frac{Y}{1 - Y}\right) = n_H(\log[B]) - \log(K_D)} \tag{16}$$

where Y is the fractional saturation of sites (occupied sites/total sites) and n_H is the Hill coefficient.

Equation 16 defines a line with $\log(Y/[1 - Y])$ plotted on the Y axis, $\log[B]$ plotted on the X axis, and a slope equal to n_H. The term n_H is a nonintegral parameter that relates to the cooperativity in the system; note that n_H is not the number of B molecules that bind to A. If n_H for a data set is greater than 1, the system exhibits positive cooperativity. Values of n_H equal to or less than 1 have no meaning without additional information regarding the affinity constants for B binding each site on A. In the Hill equation the Y intercept equals $-\log K_D$. A K_D derived from the Y intercept of a Hill plot is meaningless for biological interactions that do not occur with perfect cooperativity; it is not the K_D for any single B molecule associating with A.

***Fractional saturation* (Y).** Importantly, Y is not the same as the fraction bound that is described in Chapter 2 and earlier in this chapter. Rather, Y is the number of occupied sites divided by the total number of sites. To understand Y, consider a molecule A that has two sites for binding molecule B. Here, the fractional saturation of the two sites on A at any concentration of B is:

$$Y = \frac{[AB]+2[AB_2]}{2[A]_{Total}} \quad (17)$$

Each molecule of A has two sites for B; hence, the total number of sites available to bind B molecules equals $2[A]_{Total}$. The number of sites occupied by B includes the complexes that are singly occupied ([AB]) plus two times the complexes that are doubly occupied ($2[AB_2]$). Importantly, the fractional saturation considers singly occupied complexes (AB) even though, according to the assumption of perfect cooperativity under which the Hill equation is derived, singly occupied complexes would not exist.

Hill plots. The Hill equation defines a line with log(Y/[1 – Y]) plotted on the Y axis and log[B] plotted on the X axis. The slope is the Hill coefficient n_H and the data plotted in this manner are referred to as a Hill plot. Figure 3.5 illustrates how a Hill plot can be used to evaluate the cooperative binding of two B molecules to A (A + 2B ↔ AB_2) in both a perfectly cooperative system and a highly cooperative system. The dashed line in the figure is straight and results from perfect positive cooperativity (i.e., singly occupied molecules of A (AB complexes) are not found at equilibrium at any concentration of B). The slope of the line, and therefore the Hill coefficient, is 2. (In this hypothetical case, n_H is in fact equal to the number of binding sites for B, because the system is perfectly cooperative.) The solid curve in Figure 3.5 results from a highly cooperative (but not perfectly cooperative) system that might actually exist in biology, where α is equal to 0.02. (Recall that a value of α below 1 reflects positive cooperativity; for a complete description of α and how it relates to cooperativity, see the preceding sections). This more realistic curve demonstrates an important aspect of Hill plots. Namely, when real experimental data are plotted over a wide range of B concentrations,

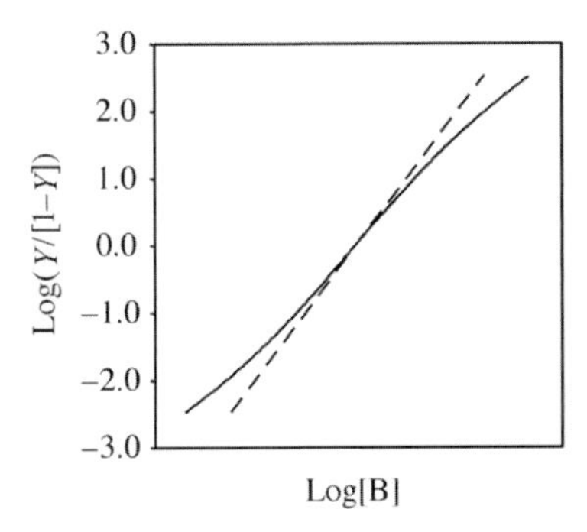

FIGURE 3.5. Hill plot of two data sets showing positive cooperativity in the system described by A + 2B ↔ AB_2. The *straight dashed line* represents a system with perfect cooperativity, unlikely to exist in biology. The *solid curve* represents a highly cooperative system that could exist in biology.

the points do not necessarily fit a straight line. Rather, at low and high concentrations of B, two asymptotes are formed with slopes near 1. We next describe how to determine the Hill coefficient from a curve such as this one.

Experimental Considerations

Measuring fractional saturation (Y). Typically, the fractional saturation of sites on A is measured when the concentration of A is held constant and well below the K_D for association with B at any of its sites. B is titrated and the fractional saturation is determined at each concentration of B. As will become clear later in this chapter, only concentrations of B that ultimately lead to a fractional saturation between 0.1 and 0.9 will be useful for determining the Hill coefficient. Therefore, a pilot experiment may be required to determine the optimal range over which B should be titrated.

There are two approaches to measuring fractional saturation depending on whether A or B is labeled. We first consider the situation in which A is labeled. Remember that fractional saturation describes the ratio of occupied sites to total sites. Therefore, if A is labeled, a method must be established to separate all the possible states of bound A in the reaction. For example, if A has two sites for binding B, then the concentrations of unoccupied A, singly occupied A (AB), and fully occupied A (AB_2) must all be measured at equilibrium. Once these values are measured, the fractional saturation is calculated for every concentration of B using Equation 17. A major disadvantage of this approach is that it is often difficult to separate and identify all of the different states of A. For this reason, researchers often label B and use the following method to determine fractional saturation.

When B is labeled, the different bound states of A need not be isolated (e.g., in the two-site model, AB and AB_2 do not need to be separately quantitated) and the equation used to calculate fractional saturation is:

$$Y = \frac{{}_{\text{Bound}}\text{B}}{{}_{\text{Bound}}B_{\text{max}}} \tag{18}$$

After reactions reach equilibrium, all complexes containing both A and B are separated from free B, and the total amount of B bound in complexes is quantitated at each concentration of B (${}_{\text{Bound}}\text{B}$ in Equation 18). To calculate the fractional saturation, one additional piece of data is required: the maximum amount of B that can possibly bind A (i.e., the amount of bound B when all sites on all molecules of A in the reaction are occupied). This value (${}_{\text{Bound}}B_{\text{max}}$ in Equation 18) is obtained from control reactions performed at a very high concentration of B. Fractional saturation is then calculated for every concentration of B using Equation 18.

Creating Hill plots. At each concentration of free B, $\log(Y/[1 - Y])$ and $\log[\text{B}]$ are calculated and plotted on the Y axis and X axis, respectively (this is different than plotting raw data on a logarithmic axis). If A was held constant at a concentration much lower than the K_D for B binding to any site on A, then the total concentration of B added to the reactions can be used to approximate the free concentration of B. To determine the Hill coefficient, only data that fall within the range where $-0.9 < \log(Y/[1 - Y]) < 0.9$ should be plotted and fit with a line. Within this range, the Hill coefficient (n_H) is equal to the slope of the line resulting from linear regression. Outside of this range, the slope of a line put through the data will approach 1. Hence, curve fitting over a broad range of B concentrations will give an inaccurate Hill coefficient.

Figure 3.6 demonstrates how to create an accurate Hill plot (see also Simulation S3-2). The data points were generated for the reaction $A + 2B \leftrightarrow AB_2$ in which the K_D values for B binding to each site on A were both 30 nM and α was 0.1, indicating positive cooperativity. (For a complete discussion of α values and cooperativity, see the previous sections in this chapter.) A broad range of points was plotted and divided into three groups to illustrate both the useful range of data (circles) and the nonlinearity of the entire curve that occurs when datapoints outside this range are considered. For concentrations of B resulting in Y values between 0.1 and 0.9 (–0.9 and 0.9 on the Y axis), the data (circles) are nicely fit by a line with a slope of 1.5; recall that the slope is the Hill coefficient. The data outside of this range (squares) were separately fit using linear regression, yielding slopes of 1.1. Because the Hill coefficient of 1.5 is greater than 1, there is positive cooperativity in this system. Importantly, the Hill coefficient neither equals nor relates to the number of B-binding sites on A nor the cooperativity factor α.

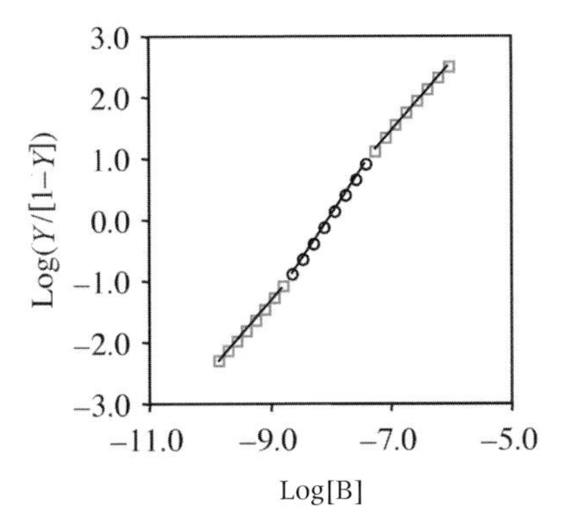

FIGURE 3.6. Hill plot showing fractional saturation over a broad range of B concentrations. Data shown as circles fall in the useful range for measuring the Hill coefficient and the line through these points has a slope of 1.5.

TABLE 3.2. Cooperativity when $K_{D1} = K_{D2}$

Cooperativity	K_{D1}	K_{D2}	α	n_H
Positive	30 nM	30 nM	0.1	1.5
None	30 nM	30 nM	1	1
Negative	30 nM	30 nM	10	0.6

Interpreting the Hill coefficient when all sites on A have the same inherent affinity for binding B. In practice, there are only a very limited number of biological systems for which it is possible, using Hill plots, to conclude that a system is noncooperative or negatively cooperative. These are systems in which all binding sites on A have identical affinities for binding a molecule of B in the absence of other bound B molecules. Most biological systems do not fall into this category. Moreover, it is difficult to independently measure the K_D values for B binding to each unique site on A; therefore, determining if all binding sites on A have equal affinities for B is often not possible. In the unique situations in which it is known that all binding sites on A have equal affinities for B, the n_H derived from a Hill plot can unambiguously be used to determine whether B molecules bind A with negative cooperativity ($n_H < 1$), positive cooperativity ($n_H > 1$), or no cooperativity ($n_H = 1$). This is illustrated in Table 3.2 and Figure 3.7, which show, respectively, the Hill coefficients and Hill plots obtained from data that were generated using the K_D and α values in the table. Because the K_D values for each site on A are equal (30 nM), the values of n_H reflect the cooperativity in the system as defined by α.

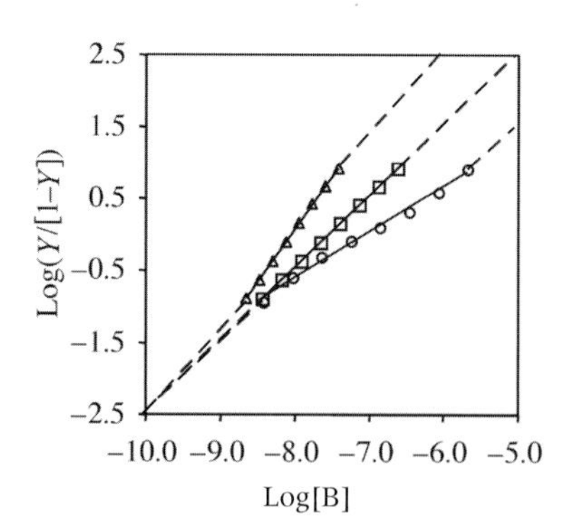

FIGURE 3.7. Hill plot that was generated using the K_D and α values shown in Table 3.2. The *triangles* define a line with a slope of 1.5, indicating positive cooperativity. The *squares* define a line with a slope of 1, indicating no cooperativity. The *circles* define a line with a slope of 0.6, indicating negative cooperativity. The *dashed lines* represent data that fall outside of the region of the curve useful for determining the Hill coefficient.

TABLE 3.3. Cooperativity when $K_{D1} \neq K_{D2}$

Cooperativity	K_{D1}	K_{D2}	α	n_H
None	1 nM	1 nM	1	1
Positive	1 nM	1 nM	**0.1**	1.4
None	1 nM	**10 nM**	1	0.8
None	1 nM	**0.1 nM**	1	0.8
Positive	1 nM	**100 nM**	**0.1**	0.8

Interpreting the Hill coefficient when all sites on A do not have the same inherent affinity for binding B. This is the situation that typically occurs in biology. In such a system, only one experimental outcome from a Hill plot has meaning: If the Hill coefficient (slope) is greater than 1, the system has positive cooperativity. Slopes equal to or less than 1 do not allow conclusions to be made regarding cooperativity. This is because it is possible to obtain a slope equal to or less than 1 for systems with positive, negative, or no cooperativity.

Table 3.3 illustrates how n_H changes when the K_D and α values change for a system in which two molecules of B bind A. The K_D and α values in each row were used to generate Hill plots, which resulted in the Hill coefficients shown in Table 3.3. The first row represents a noncooperative ($\alpha = 1$) two-site system in which the K_D values for B binding to each site are equal. In the second through fifth rows, changes in K_{D2} and α were made as highlighted in bold. As shown in the second row, if α is decreased to 0.1 (indicating positive cooperativity), the Hill coefficient increases to 1.4, consistent with the statement that Hill coefficients greater than 1 reveal positive cooperativity. The third and fourth rows demonstrate that when one of the K_D values is changed (either increased or decreased), the Hill coefficient becomes less than 1, even though the system is noncooperative ($\alpha = 1$). Moreover, the fifth row shows that when the two binding sites have different K_D values, a Hill coefficient less than 1 (suggesting negative cooperativity) can be obtained, even in a system that is positively cooperative ($\alpha < 1$).

Key points: Assessing cooperativity using Hill plots

$n_H > 1$, positive cooperativity

$n_H \leq 1$, cooperativity unknown (unless $K_{D1} = K_{D2} = K_{Dn}$)

$n_H \neq$ number of B binding sites

K_D obtained from the Y intercept is meaningless

Literature Example 3.1. Assessing cooperative DNA binding using a Hill plot.
(Bruck I., Woodgate R., McEntee K., and Goodman M.F. 1996. J. Biol. Chem. 271: 10767–10774.) Figure reprinted, with permission, from Bruck et al. (1996). (© American Society for Biochemistry and Molecular Biology)

The UmuD′C complex functions in the SOS response to DNA damage in *Escherichia coli.* It interacts with single-stranded DNA (ssDNA) and induces mutagenesis. To determine whether UmuD¢C complexes bind ssDNA with positive cooperativity, a mobility shift assay was performed using native agarose gels. The UmuD′C complex (molecule B in Equation 14) was titrated from 25 nM to 450 nM into reactions containing 5.4 nM ssDNA, which was 600 nucleotides in length (molecule A in Equation 14). The reactions were incubated for 20 minutes and then run on a native agarose gel to separate bound protein from free protein. The protein in the gel was transferred to a nitrocellulose membrane, and the free and bound proteins were detected by western blotting. Bands were quantitated by densitometry while ensuring that the signals were in the linear range for detection. Fractional saturation values were calculated at every concentration of added UmuD′C using Equation 18.

The resulting Hill plot is shown. (Note that the units of the *X* axis are log[nM].) Data were plotted over the range of –0.9 to 0.5 on the *Y* axis. The slope of the line (and thus the Hill coefficient) was 3, showing that the interaction of UmuD′C with ssDNA is positively cooperative. Interestingly, there was no evidence for cooperative interaction of UmuD′C with ssDNA 80 nucleotides in length.

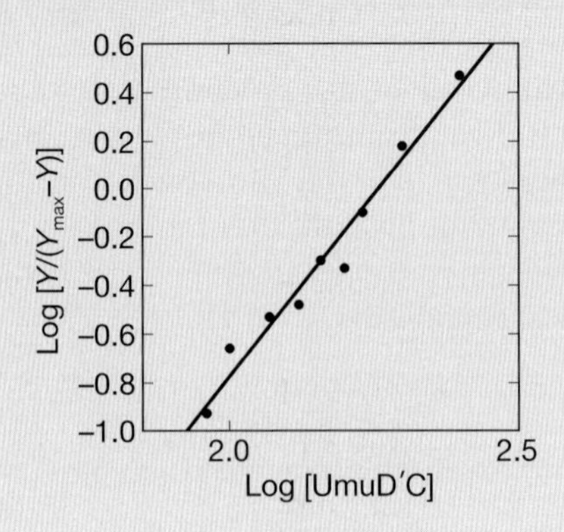

CHAPTER 4

Rate Constants for Binding and Dissociation

THE PRIMARY OBJECTIVE OF QUANTITATIVE kinetic studies is to measure the rate constants for reactions. Rate constants are numbers that relate the rate at which a reaction occurs to the concentration of the biomolecules involved in the reaction. Therefore, rate constants provide quantitative meaning to phrases such as "fast binding" and "slow dissociation." Beyond placing a quantitative value on how fast or slow reactions occur, kinetic studies have the potential to provide insight into the mechanisms of reactions, as well as the temporal regulation of biological processes in cells.

Topics covered:

- General rate theory
- Dissociation of two biomolecules
- Association of two biomolecules

In this chapter, we consider measuring rate constants for bimolecular binding reactions; kinetic studies of more complex binding reactions are considered in Chapters 5 and 6, respectively. The intent of this chapter is to provide enough theory to understand quantitative kinetic experiments and how to determine rate constants for most two-component biological interactions. Examples from the literature are provided to demonstrate the practical aspects of measuring rate constants. Key equations are displayed in boxes and derived in Appendix 1. In addition, computer simulations are provided on the Web site (http://kinetics.cshl.edu) that can be manipulated to help visualize concepts presented in the chapter.

TERMS AND PRINCIPLES

Single-step bimolecular binding reactions can be described as $A + B \leftrightarrow AB$. The kinetics of both association and dissociation can be studied.

Kinetic studies measure changes in the concentrations of reactants and/or products that occur in reactions over time. To study the kinetics of a binding reaction, the equilibrium between free and bound species is perturbed and changes in their concentrations are measured over time as the reaction approaches a new state of equilibrium.

Rate of a binding reaction describes the change in the concentration of a molecule or complex per change in time. The units for the rate of a binding reaction are $\text{M}t^{-1}$ (M/t), where M is molar and t is time (e.g., seconds or minutes).

Rate constants for binding reactions are numeric representations of the time that it takes for molecules to associate or dissociate. In general, rate = (rate constant)(concentrations).

Association rate constant **(k_1)** relates the rate at which biomolecules associate to the concentrations of the molecules. For a bimolecular interaction, the association can be depicted as A + B → AB. In this case, the rate of association is equal to k_1[A][B] and the rate constant k_1 has units of $\text{M}^{-1}t^{-1}$, where M is molar and t is time (e.g., seconds or minutes). In general, the larger the value of k_1, the faster the inherent rate of association between A and B.

Dissociation rate constant **(k_{-1})** relates the rate at which biomolecules in a complex dissociate to the concentration of the complex. For a bimolecular interaction, the dissociation can be depicted as AB → A + B. In this case, the rate of dissociation is equal to k_{-1}[AB] and the rate constant k_{-1} has units of t^{-1}, where t is time (e.g., seconds or minutes). In general, the larger the value of k_{-1}, the faster the inherent rate of dissociation.

Observed rate constants **(k_{obs})**, as explained in this chapter, are experimentally measured values that describe the formation of a complex over time. Observed rate constants mathematically relate to the actual rate constants for association (k_1) and dissociation (k_{-1}) and have the units of t^{-1}, where t is time (e.g., seconds or minutes).

Half-time for an association reaction is the time at which complex formation is 50% complete after the interacting biomolecules are combined. For a dissociation reaction, half-time is the time at which 50% of the complex remains after equilibrium has been perturbed and dissociation begins to be monitored.

Diffusion limited rate refers to binding reactions in which the rate of diffusion in solution limits the forward rate of association of two biomolecules. For a single-step reaction, it is impossible for two molecules to associate faster than they encounter each other in solution; therefore, the rate of the reaction is diffusion limited. For reactions that are diffusion limited, the association rate constant (k_1) will be in the range of 10^8 to 10^9

$M^{-1}s^{-1}$ (under common biological experimental conditions). Note that the rate of dissociation is not limited by the rate of diffusion.

GENERAL RATE THEORY

The following discussion and equations are applicable to many biological reactions. Their application to bimolecular binding reactions constitute the remainder of this chapter.

Rates and Rate Constants

Consider the expression:

$$\text{reactant} \underset{k_r}{\overset{k_f}{\rightleftharpoons}} \text{product} \tag{1}$$

The top arrow indicates the forward reaction, which is governed by a forward rate constant (k_f) and the bottom arrow indicates the reverse reaction, which is governed by a reverse rate constant (k_r). We use the general terms k_f and k_r for rate constants in this section as a means of demonstrating general rate theory. The rate of a reaction is defined as the change in the concentrations of reactants and products over time as the reaction approaches equilibrium. When considered independently, the rate of the forward reaction relates to the forward rate constant and concentration of reactant by the following equation:

$$\text{forward rate} = k_f[\text{reactant}] \tag{2}$$

Similarly, the rate of the reverse reaction relates to the reverse rate constant and concentration of product by the following equation:

$$\text{reverse rate} = k_r[\text{product}] \tag{3}$$

The overall rate of product accumulation as equilibrium is approached must take into account both the forward and reverse rates shown above. The rate of product accumulation is equal to the forward rate minus the reverse rate, as described by the following equation:

$$\text{rate of product accumulation} = k_f[\text{reactant}] - k_r[\text{product}] \tag{4}$$

Similarly, the rate of product decay is equal to the reverse rate minus the forward rate:

$$\text{rate of product decay} = k_r[\text{product}] - k_f[\text{reactant}] \tag{5}$$

It is important to remember that when a reaction is at equilibrium, there is no net change in reactants and products (i.e., at equilibrium k_f[reactant] = k_r[product]).

Equations 4 and 5 can be simplified under some conditions. For example, when k_f is far greater than k_r, the reaction will reach an equilibrium condition where [product] is far greater than [reactant]. In this situation, when the rate of product accumulation is measured after initiating a reaction, the rate of product decay is negligible, that is, the rate of product accumulation approximates the forward rate of the reaction. Therefore, the k_r[product] term in Equation 4 is negligibly small and the equation simplifies to Equation 6:

$$\text{rate of product accumulation} = k_f[\text{reactant}] \tag{6}$$

Similarly, the rate of product decay can be experimentally measured under conditions where the forward reaction is inhibited. In this situation, Equation 5 simplifies and the rate of product decay equals the reverse rate of the reaction.

$$\text{rate of product decay} = k_r[\text{product}] \tag{7}$$

Measuring Rate Constants: Exponential Expressions

The primary goal of quantitative kinetic experiments is to determine the rate constant for the reaction of interest, which typically involves measuring the accumulation of product over time or the disappearance of product over time (forward rate or reverse rate, respectively) and using the appropriate equations to determine the rate constant. The concentration of product increases or decreases over time in an exponential manner; therefore, the most useful equations to describe the reaction rates contain an exponential term. Equations 6 and 7, which describe the forward and reverse rates for a reaction, can be mathematically transformed into equations containing exponential terms that are experimentally useful for determining rate constants. For the forward reaction:

$$\frac{[\text{product}]_t}{[\text{product}]_f} = 1 - e^{-k_f t} \tag{8}$$

Here, $[\text{product}]_t$ refers to the concentration of product at each time point (t) after initiating the reaction and $[\text{product}]_f$ refers to the final (f) concentration of product formed at the end of the reaction. Therefore, $[\text{product}]_t/[\text{product}]_f$ is the fraction of product formed at any time point. k_f is the forward rate constant for the reaction and has units of t^{-1} (e.g., s^{-1}).

For the reverse reaction, the decay of product back to reactant over time is monitored:

$$\frac{[\text{product}]_t}{[\text{product}]_i} = e^{-k_r t} \tag{9}$$

Here, $[\text{product}]_i$ refers to the initial (*i*) concentration of product present at the beginning of the reaction and $[\text{product}]_t$ refers to the concentration of product at each time point (*t*) after beginning to monitor the reverse reaction. Therefore, $[\text{product}]_t/[\text{product}]_i$ is the fraction of product remaining at any time point during the reverse reaction. k_r is the reverse rate constant for the reaction and has units of t^{-1} (e.g., s^{-1}).

Plotting Rate Data

To determine the forward or reverse rate constant, data are plotted with the fraction of product formed (forward reaction) or the fraction of product remaining (reverse reaction) on the *Y* axis versus time on the *X* axis and fit with Equations 8 and 9, respectively. The plot in Figure 4.1 shows a forward reaction curve generated with Equation 8, and the plot in Figure 4.2 shows a reverse reaction curve generated with Equation 9. Both plots illustrate the exponential progress of the reactions. In both cases, the rate constant is equivalent to the inverse of the time point at which the reaction is 63% complete. This is because when the rate constant (k) equals $1/t$, $e^{-kt} = 0.37$. For the forward reaction, this point occurs

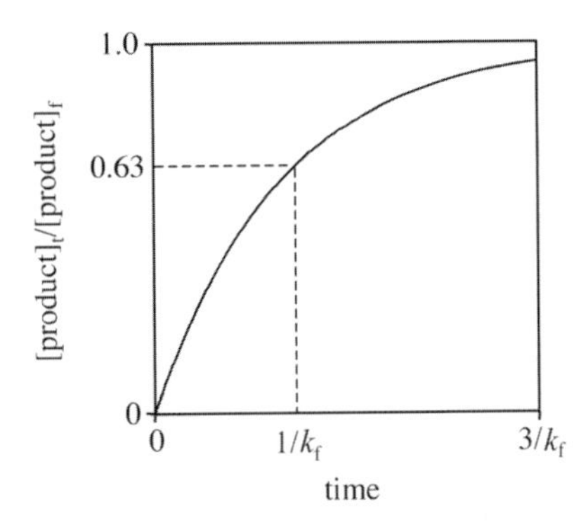

FIGURE 4.1. Curve for a forward reaction defined by Equation 8. The progress curve for any single exponential forward reaction will have this shape over the range of time from zero to $3/k_f$ (plotted on the *X* axis). The *dashed lines* indicate the time at which the reaction is 63% complete, which is equal to $1/k_f$.

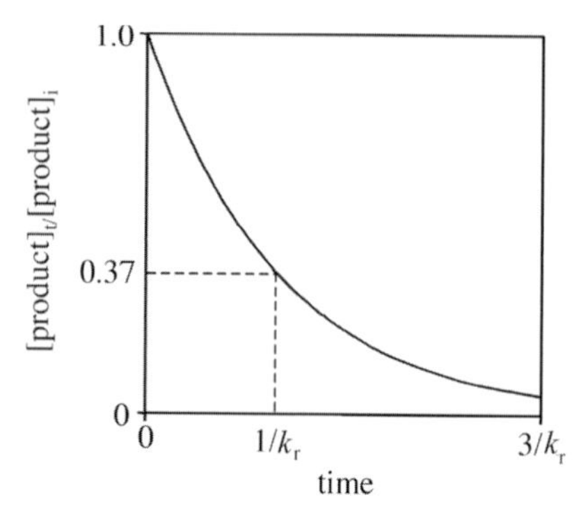

FIGURE 4.2. Curve for a reverse reaction defined by Equation 9. The progress curve for any single exponential reverse reaction will have this shape over the range of time from zero to $3/k_f$ (plotted on the X axis). The *dashed lines* indicate the time at which the reaction is 63% complete (0.37 on the Y axis), which is equal to $1/k_r$.

when the fraction of product formed equals 63% ($[\text{product}]_t/[\text{product}]_f = 1 - 0.37 = 0.63$), and for the reverse reaction, this point occurs when the fraction of product remaining equals 37% ($[\text{product}]_t/[\text{product}]_i = 0.37$).

Reaction Order

The order of a reaction is equal to the number of reactants whose concentrations affect the rate of the reaction. Reactions of zero, first, second, and third orders are known to occur in nature; however, in this chapter (and throughout the book), we exclusively consider first- and second-order reactions.

The rate of a first-order reaction depends on the concentration of only a single reactant, for example, $A \rightarrow B$ or $AB \rightarrow A + B$. In both cases, there is only a single reactant on the left side of the arrow (A or AB). For first-order reactions, the rate of the reaction is equal to the molar concentration of the reactant ([A] or [AB] in the examples) multiplied by the first-order rate constant (units of t^{-1}). Hence, rate = k[reactant].

The rate of a second-order reaction depends on the concentrations of two reactants, for example, $A + B \rightarrow AB$ or $A + A \rightarrow AA$. In both cases, two reactants are on the left side of the arrow. For second-order reactions, the rate of the reaction is equal to the product of the molar concentration of the reactants ([A][B] or [A][A] in the examples) multiplied by the second-order rate constant (units of $\text{M}^{-1}t^{-1}$). Hence, rate = k[reactant][reactant].

DISSOCIATION OF TWO BIOMOLECULES (AB → A + B)

In this section, we describe approaches to determining the dissociation rate constant for bimolecular binding reactions. We chose to discuss the kinetics of dissociation before considering association for two reasons. First, the theory and experimental principles behind measuring dissociation rate constants follow directly from the general rate theory described above. Understanding how dissociation rate constants are measured will provide a useful platform to explain how to measure association rate constants, which requires a unique extension of the general rate theory discussed above. Second, it is often experimentally easier to measure dissociation rate constants than it is to measure association rate constants. This is because for most biological binding reactions, dissociation occurs much more slowly than association. Dissociation often occurs over tens of seconds to tens of minutes, hence sampling can be performed manually. In contrast, many association reactions are complete within seconds or less. Thus, it is not always possible to measure association rate constants via manual sampling. In these situations, automated assays are required to measure accurate association rate constants.

Theory and Equations

Bimolecular binding interactions can be described by the expression:

$$\mathrm{A} + \mathrm{B} \underset{k_{-1}}{\overset{k_1}{\rightleftharpoons}} \mathrm{AB} \tag{10}$$

where A and B represent two interacting biomolecules. The bottom arrow indicates the dissociation of the complex, which is governed by a dissociation rate constant (k_{-1}) with the units of t^{-1}.

Studying dissociation in the absence of association. The change in AB complex over time is equal to the rate of AB formation minus the rate of AB dissociation:

$$\frac{\text{change in [AB]}}{\text{change in time}} = k_1[\mathrm{A}][\mathrm{B}] - k_{-1}[\mathrm{AB}] \tag{11}$$

Here, [A], [B], and [AB] are molar concentrations. Measuring the dissociation rate is most simply done under conditions where the association reaction does not occur (i.e., the rate of association, k_1[A][B], is zero and can be ignored). Under these conditions, the rate at which AB changes over time approximates the rate at which the AB complex dissociates and Equation 11 can be simplified to:

$$\frac{\text{change in [AB]}}{\text{change in time}} = -k_{-1}[\text{AB}] \tag{12}$$

The general approach to studying the rate of dissociation involves three steps, performed in the following order: (1) forming the AB complex, (2) changing the reaction conditions to prevent new association of A and B from occurring, and (3) measuring the change in [AB] over time. The second step is critical because eliminating the association reaction allows dissociation to be studied in the absence of reassociation, and Equation 12 can be used. Experimental approaches to block the association reaction without perturbing the AB complex are discussed later.

***Transforming Equation 12 to simplify measuring* $\mathbf{k_{-1}}$.** Ultimately, the goal is to measure the dissociation rate constant (k_{-1}), which describes the exponential decay of AB complexes. To facilitate measuring k_{-1}, Equation 12 can be mathematically transformed into the following single exponential equation (for derivation, see Appendix 1):

$$\boxed{\frac{[\text{AB}]_t}{[\text{AB}]_i} = e^{-k_{-1}t}} \tag{13}$$

$[\text{AB}]_t/[\text{AB}]_i$ is the concentration of the AB complex at each time point ($[\text{AB}]_t$) divided by the initial concentration of the AB complex ($[\text{AB}]_i$) in the reaction just before the point at which new association is blocked (t = 0). We refer to $[\text{AB}]_t/[\text{AB}]_i$ as the fraction remaining. Importantly, by definition, $[\text{AB}]_t/[\text{AB}]_i$ equals 1 at time zero (i.e., $[\text{AB}]_t = [\text{AB}]_i$ when $t = 0$).

The concentrations of A and B will affect the initial concentration of AB complex that forms ($[\text{AB}]_i$); however, the rate constant for dissociation of AB (k_{-1}) is independent of [AB], [A], and [B]. This is one reason why measuring dissociation rate constants is simpler than measuring association rate constants: The concentrations of A and B do not affect the measurement of k_{-1}. Note that there are no terms in Equation 13 that depend on the concentration of A or B.

Equation 13 describes a curve that starts at a fraction remaining of 1 and exponentially decreases to 0 as time approaches infinity (see Fig. 4.2). To determine the dissociation rate constant, the fraction remaining in the complex is plotted on the *Y* axis and time on the *X* axis. The data points can be fit with a curve defined by Equation 13 to determine k_{-1}, as explained in detail in subsequent sections.

Half time. In some situations, it may be useful to know the half-time for a dissociation reaction. This refers to the time required for half of the AB complex to dissociate after new association is blocked. The equation for the half-time can be determined using Equation 13, substituting 0.5 for $[AB]_t/[AB]_i$ and solving for time to obtain (for derivation, see Appendix 1):

$$t_{1/2} = \frac{\ln 2}{k_{-1}} = \frac{0.693}{k_{-1}} \quad (14)$$

Because $\ln(2) = 0.693$, the $t_{1/2} = 0.693/k_{-1}$ for any dissociation reaction.

Experimental Considerations

The remainder of our discussion surrounding dissociation rate constants is divided into three sections that focus on the practical aspects of experimentally measuring k_{-1}. First, we describe how to measure rate constants based on the equations presented in the theory section above. In the section "Dealing with Aberrant Decay Data," we describe equations that are useful when data do not follow an ideal decay curve that begins at 1 and levels out at 0. In the third section, "General Equations for Fitting Decay Data," we present two equations that can be used to fit any decay data. Note, however, that much of the theory behind these equations is presented in the two preceding sections.

***Developing a method to measure* $\mathbf{k_{-1}}$.** Measuring a dissociation rate constant involves three steps: allowing the maximum amount of AB complex to form given the concentrations of A and B added to the reaction, blocking new association from occurring, and monitoring the decay of the AB complex over time. Therefore, to measure k_{-1}, three experimental conditions must be considered: (1) An assay must be developed to measure AB, (2) a method must be established to block the association of A with B, and (3) a time course appropriate for the experiment must be defined. Each of these points is discussed below.

1. *Measuring AB.* An assay to accurately measure the AB complex over time is necessary. (Alternatively, an assay to measure free A or B can be used, as discussed later.) The assay must distinguish the AB complex from unbound A or B, such that the concentration or amount of AB complex can be determined at each time point. Often, one biomolecule (A or B) is labeled, for example, with radioactivity or a fluorophore. The type of assay not only depends on the nature of the interaction (e.g., protein–protein or protein–nucleic acid), but also depends on how fast dissociation occurs (e.g., seconds, minutes, or

hours). For example, some techniques such as spectroscopy allow changes in the AB complex to be monitored over time without removing samples, which is very convenient for rapid time courses. Other techniques, such as electrophoretic mobility shift assays, require that samples be removed from the reaction and analyzed at multiple time points. In this case, if the time course for dissociation is hours, a series of reactions can be initiated at different times (over hours), and at the end of the time course, all of the reactions can be loaded on a gel.

2. *Blocking the association of A with B.* Unique to measuring dissociation rate constants is the need either to block new association or to render newly formed complexes undetectable. The choice of technique for blocking the reassociation of A and B will depend on the characteristics of the molecules being studied; several commonly used methods are described below.

   a. *Dilution.* One technique to effectively block the association reaction is massive dilution. After the maximum amount of AB complex has formed, the reaction can be diluted such that the concentrations of A and B will be well below the K_D governing the AB complex. Therefore, after dilution, AB complexes that undergo dissociation will not reform because the concentrations of A and B will not be high enough to allow association to occur.

   b. *Nonspecific competitor.* The association reaction can be blocked with a nonspecific competitor molecule. High amounts of a competitor trap either free A or B and prevent AB complexes from forming. Consider the example in which A is a molecule of DNA containing a target sequence to which protein B binds. In this case, a nonspecific competitor nucleic acid such as calf thymus DNA can be used to block the association reaction. When added after AB complex formation is complete, the calf thymus DNA will sequester free B, thereby blocking the formation of new AB complexes. Importantly, the calf thymus DNA must be added in sufficient quantities to sequester B more rapidly than B can rebind A. This can be determined using control experiments in which the competitor is added to reactions before A and B are combined; if no AB complex is observed, then the competitor effectively blocks the association reaction.

   c. *Specific competitor.* A specific competitor can be used to eliminate the association reaction. Typically, unlabeled A or B is used as a specific competitor for labeled B or A, respectively. In this case, the association reaction is not formally blocked, but any new complexes that form will not be labeled and hence will not be detected. For example, if protein A is labeled, a great excess of unlabeled A (e.g., at least 100-fold) can be added to reactions after the maximum amount of AB complex has formed. The free B, released as the labeled AB complexes decay, will associate with the excess of unlabeled A in the reaction, but these new AB complexes will not be detectable. In using this technique, it is important to make sure that the concentration of unlabeled A is greater than the concentration of B in the reactions.

3. *Establishing a time course for the experiment.* Experiments to measure the dissociation rate constant are performed under conditions where time points are taken after initiating the reaction by blocking new association of A with B. The number of time points and the overall time course depend on the reaction being studied and the method being used. In general, time points should begin tenfold below the $t_{1/2}$ and end tenfold above. Using less than ideal time courses can negatively impact the accuracy of the rate constant determined. Several pilot experiments may be required to establish the ideal time course.
4. *A critical control.* It is important to ensure that the condition used to block new association of A with B does not itself affect the dissociation rate constant being measured. This potential problem can be assessed by measuring dissociation rate constants using several concentrations of competitor or points of dilution. To appreciate the importance of this problem, consider a situation in which the measured dissociation rate constant increases as the concentration of a nonspecific competitor (used to sequester free B) is increased. In this case, the competitor not only acts to sequester free B, but also might destabilize the AB complex in a concentration-dependent fashion. The dissociation rate constant measured using this competitor may not be the true k_{-1}, and another means of blocking new AB complex formation should be developed.

Concentration and sustained activity of A and B. As we discussed in the Theory and Equations section, the dissociation rate constant does not depend on the concentrations of A and B added to reactions. Be aware that if the concentrations of A and B are near or below the K_D, then the amount of AB complex may be low when the reaction starts (for an explanation of K_D, see Chapter 2). This will not affect the measured value of k_{-1}, however, measuring AB complex decay may be more difficult.

It is possible that in vitro experimental artifacts, such as loss in a molecule's activity, may occur during the long time courses often required to measure dissociation rate constants. For example, protein preparations might contain contaminating proteases capable of degrading the protein of interest over the time course of the experiment. In this case, degradation of one of the interacting biomolecules would decrease the amount of AB complex in the reaction over time. Depending on the method used to detect AB complexes, it might not be possible to distinguish between loss of AB complex due to dissociation (governed by k_{-1}) and loss of AB complex due to degrading either A or B in the complexes (governed by a rate constant distinct from k_{-1}). In this case, the measured rate constant for dissociation would be a composite of k_{-1} and another (or perhaps multiple) rate constant, causing the measured rate constant for dissociation of AB to be greater than the actual rate constant (k_{-1}). If it is suspected that in vitro artifacts are influencing the measurement of a dissociation rate constant (and the cause of the artifacts cannot be identified and eliminated), it is best to present the measured k_{-1} as an upper limit, rather than an absolute value.

Determining* k_{-1} *when* $[AB]_t/[AB]_i$ *is measured. The following approach to determining k_{-1} follows directly from the Theory and equations presented in the previous section; we discuss variations on this approach later in the chapter. At each time point, the concentration of AB is quantitated ($[AB]_t$). To obtain the fraction remaining at each time point, the $[AB]_t$ values are divided by the initial concentration of AB complex ($[AB]_i$). $[AB]_i$ is measured just before blocking new association of A with B. At time zero, $[AB]_t/[AB]_i$ will equal 1.

Data are plotted with the fraction remaining ($[AB]_t/[AB]_i$) on the *Y* axis and time on the *X* axis. The plot in Figure 4.3 depicts a perfect decay curve that begins at 1 and approaches 0. The k_{-1} is 0.05 min^{-1}. (Note that the units are in min^{-1} because time was plotted in minutes on the *X* axis.)

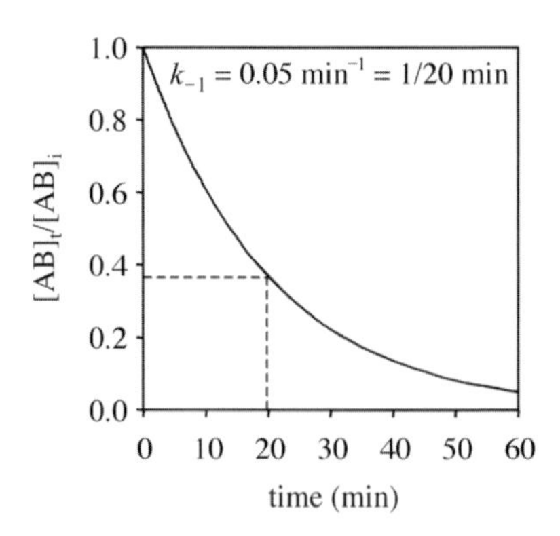

FIGURE 4.3. Decay curve for the dissociation of a bimolecular complex. The *dashed lines* indicate the time at which the reaction is 63% complete (0.37 on the *Y* axis), which is equal to $1/k_{-1}$.

k_{-1} can be found by drawing a horizontal line from 0.37 on the *Y* axis to the curve and then drawing a vertical line from this point down to the *X* axis. The k_{-1} is the inverse of the time point at which the vertical line crosses the *X* axis. Alternatively, k_{-1} can be determined using a curve-fitting program and Equation 13 (as described in Chapter 7). The half-time in this example (i.e., the time at which 50% of the AB complex has dissociated) is 13.9 minutes ($t_{1/2}$ = 0.693/0.05).

Determining* $\mathbf{k_{-1}}$ *by measuring AB: Often, it is most convenient to simply quantitate the amount of AB complex remaining over time (AB_t) as opposed to determining the concentration of AB complex at each time point ($[AB]_t$) and dividing by $[AB]_i$. When AB_t is measured instead of $[AB]_t/[AB]_i$, the following equation should be used to determine the k_{-1} (for derivation, see Appendix 1):

$$AB_t = AB_i\ (e^{-k_{-1}t}) \tag{15}$$

Here, AB_t is not a concentration, nor is it a fraction; it is the amount of AB complex at each time point expressed in a manner reflecting the method of quantitation used in the experiment. For example, AB_t could be expressed in phosphorimager units, fluorescence, scintillation counts, etc. AB_i is the amount of AB complex that is present before new association is blocked (at t = 0), and it has the same units as AB_t.

The use of Equation 15 is illustrated by the example shown in Figure 4.4. Here, B was fluorescently labeled, and the AB complex was quantitated by fluorescence spectroscopy. Fluorescence units were plotted on the *Y* axis and time in minutes on the *X* axis. The curve begins at a point

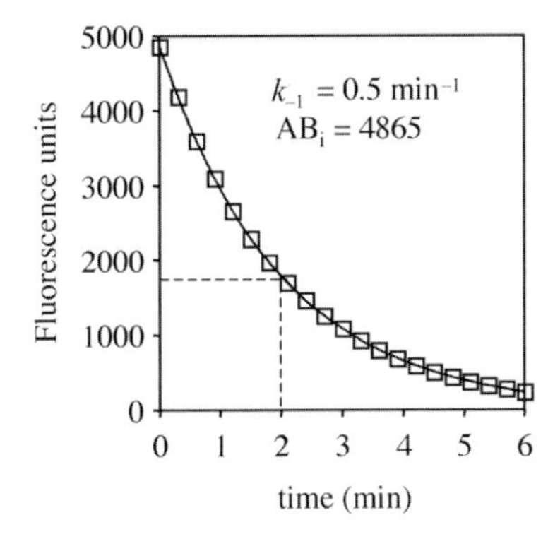

FIGURE 4.4. Decay curve for the dissociation of AB, obtained by plotting fluorescence units (i.e., the amount of AB) versus time. The k_{-1} and AB_i used to generate the data points and curve are shown.

defined by the measured value of AB_i, which is plotted at $t = 0$. As time increases, the fluorescence units approach 0. The values of k_{-1} and AB_i can be simultaneously obtained by fitting the data to Equation 15. In the example shown in Figure 4.4, AB_i is 4865 and k_{-1} is 0.5 min^{-1}, which is equal to the inverse of the time point at which the fluorescence units are 37% of AB_i (1800 in this case).

Determining* k_{-1} *by watching A or B appear. Conventional methods to determine the dissociation rate constant involve monitoring the AB complex over time. Alternatively, it may be more useful to monitor the appearance of free A or free B as the AB complex dissociates. When the accumulation of free B is measured over time, the following equation should be used to fit the data:

$$B_t = B_i + (B_f - B_i)\ (1 - e^{-k_{-1}t}) \qquad (16)$$

where B_t is the amount of free B at any time point, B_i (i for initial) is the amount of free B present in the reaction at $t = 0$ (before adding competitor), and B_f (f for final) is the amount of free B present at long times (as the plateau is approached). If the dissociation reaction goes to completion, B_f will equal the amount of B added to the reaction. Figure 4.5 shows an example plot where the accumulation of free B (expressed in phosphorimager units) was measured over time after adding a competitor that bound and sequestered free A. In this example, free B was present initially in the reaction, so the curve starts at a value of 4000 (B_i). Over time, the amount of free B increases and approaches a value of 40,000 (B_f).

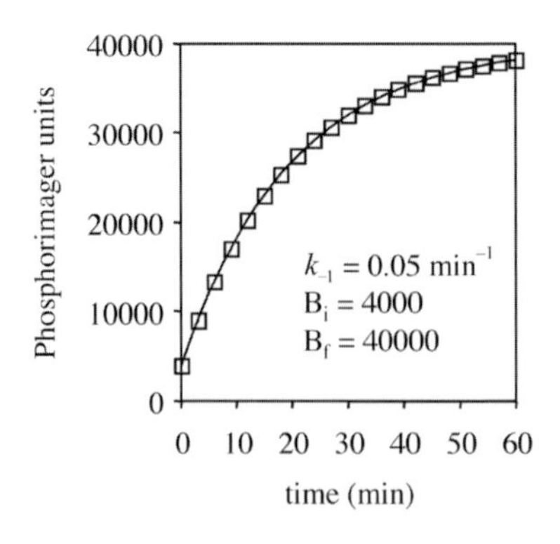

FIGURE 4.5. Time course for the dissociation of AB, obtained by plotting the accumulation of free B (plotted in phosphorimager units) over time. The k_{-1}, B_i, and B_f used to generate the data points and curve are shown.

Literature Example 4.1. Determining the dissociation rate constant for a protein–protein interaction by monitoring the appearance of free protein.
(Abel R.L., Haigis M.C., Park C., and Raines R.T. 2002. Anal. Biochem. 306: 100–107.)
Figure reprinted, with permission of Elsevier, from Abel et al. (2002).

Ribonuclease A (RNase A) and ribonuclease inhibitor (RI) protein form a very tight protein–protein complex. With the goal of characterizing RNase A variants with cytotoxic activities, these authors describe a new assay to monitor complexes between RI and RNase A. The assay used an RNase A variant (G88R RNase A) labeled at a specific residue with fluorescein. When complexed with RI, the fluorescence intensity of the RNase A variant decreased. Therefore, as the complex dissociated, unbound G88R RNase A was released into solution, causing the fluorescence to increase.

To determine the dissociation rate constant for the complex between RI and fluorescein-G88R RNase A, an equimolar solution of the two proteins was first allowed to reach equilibrium. The forward reaction was then blocked by the addition of human angiogenin, which binds RI and served as a competitor for the RI released into solution as the complex dissociated. The appearance of unbound fluorescein-G88R RNase A was monitored over time by fluorescence spectroscopy to obtain the curve shown in the figure below.

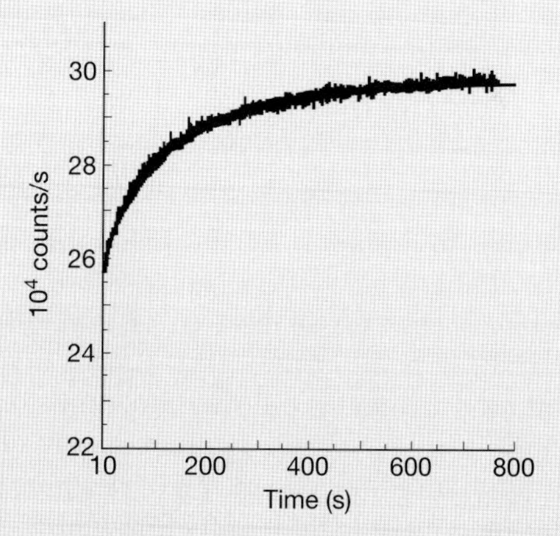

The data points were fit with the following equation (analogous to Equation 16):

$$\text{fluor} = \text{fluor}_{t=0} + (\text{fluor}_{t=\infty} - \text{fluor}_{t=0})\ (1 - e^{-k_{-1}t})$$

where k_{-1} is the dissociation rate constant, which was determined to be $7.5 \pm 0.4 \times 10^{-3}\ \text{s}^{-1}$.

Dealing with Aberrant Decay Data

Many experimental curves used to measure dissociation rate constants exhibit the characteristics described above; however, there are times when the experimental data do not simulate a perfect exponential decay curve. For example, data can show an initial burst of rapid decay followed by a slower dissociation of the remaining AB complex, which can be caused by rapid dissociation of nonspecific AB complexes. Alternatively, a decay curve can level out at a value greater than zero, which can be caused by the presence of aggregates containing A and B that do not dissociate over reasonable time courses. It is best to eliminate these problems before measuring the dissociation rate of relevant AB complexes; however, this is not always possible. This section describes methods to determine dissociation rate constants when data do not fit a perfect rate curve.

What to do when $[AB]_t/[AB]_i$ decreases rapidly at first, then slowly thereafter. It is possible that a heterogeneous mixture of AB complexes exists in solution, including the biologically relevant stable complexes and other complexes in less stable conformations. The former decays with a measurable dissociation rate constant, whereas the latter decays rapidly, perhaps within seconds. Often, the rapidly dissociating complexes result from nonspecific associations of A and B that cannot be readily distinguished from the relevant AB complexes. The following equation can be used in this situation:

$$\frac{[AB]_t}{[AB]_i} = f_{max}\,(e^{-k_{-1}t}) \tag{17}$$

Equation 17 is similar to Equation 13, but it contains an additional variable, f_{max}, which defines the starting point of the true decay curve. Therefore, Equation 17 describes a decay curve that does not have to start at 1. Rather, the starting point of the curve (f_{max}) is determined by fitting the data with Equation 17. A curve-fitting program can be used to simultaneously solve for k_{-1} and f_{max} when the fraction remaining ($[AB]_t/[AB]_i$) is plotted on the Y axis and time on the X axis. Importantly, the zero time point (where the fraction remaining is 1 by definition) should not be included in the data set used for the computer fitting.

The plot in Figure 4.6 was generated using Equation 17, a k_{-1} value of 0.05 min^{-1}, and an f_{max} value of 0.6. In this example, only 60% of the AB complexes remain by the first time point after blocking new association. This situation is indicative of a rapid decay of nonspecific complexes. Therefore, to obtain a curve that best fits the data, the fraction remaining at $t = 0$ (which by definition is equal to 1 and is obtained by measuring $[AB]_i$ just before blocking new association) is not included in the data set fit with Equation 17, as shown by the solid curve in Figure 4.6. The dashed

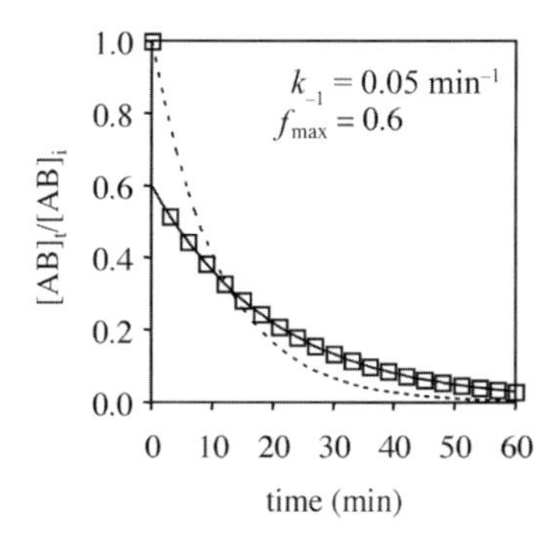

FIGURE 4.6. Decay curve for the dissociation of AB under conditions where a portion of the AB complex dissociates rapidly and the remainder dissociates with a slower and measurable rate. The k_{-1} and f_{max} used along with Equation 17 to generate the data points and the solid curve are shown. The *dashed curve* demonstrates the poor fit of the data obtained using Equation 13 (the k_{-1} determined using Equation 13 is 0.09 min^{-1}).

Literature Example 4.2. Investigating the kinetics by which the 30S ribosomal subunit assembles. *(Recht M.I. and Williamson J.R. 2001. J. Mol. Biol. 313: 35–48.)*
Figures reprinted, with permission of Elsevier, from Recht and Williamson (2001).

The 30S subunit of the bacterial ribosome is composed of one molecule of 16S rRNA and a multitude of ribosomal proteins, which are thought to bind to the RNA in a specific order during ribosome assembly. For example, binding of the S6 and S18 proteins is dependent on prior binding of the S15 protein to the 16S rRNA. These authors used kinetic experiments to investigate the molecular basis for this ordered assembly. They found that the S6 and S18 proteins form a stable heterodimer, which then binds to the S15-rRNA complex. As part of their studies, the authors included experiments to determine the rate constant for dissociation of the S6:S18 heterodimer from the S15-rRNA complex. Although there were four components in this system, a single dissociation event was monitored in isolation, and therefore the reaction could be modeled as a bimolecular dissociation.

Electrophoretic mobility shift assays were used to monitor the dissociation of S6:S18 heterodimers from the S15-rRNA complex. The four-component complex was formed by incubating a 10-nM S6:S18 heterodimer with 100-nM S15 and 2-nM ^{32}P-labeled rRNA for 1 hour to reach equilibrium. Note that S6:S18 does not bind to the rRNA in the absence of S15. In addition, the high concentration of S15 ensured that all of the rRNA in the reaction would be occupied by S15. After allowing the reactions to reach equilibrium, they were diluted 100-fold, which blocked new associ-

(Continued on following page)

Literature Example 4.2. *(Continued)*

ation because the concentrations were well below the K_D for interaction. After dilution, portions of the reaction were removed at various time points and resolved on a native polyacrylamide gel.

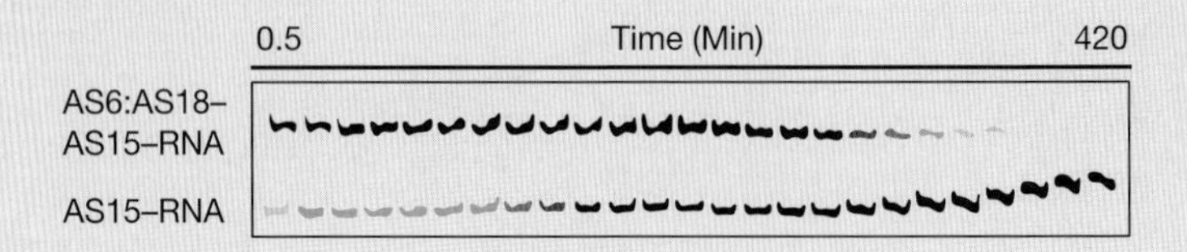

The top band is the four-component complex and the bottom band is the S15-RNA complex after the S6:S18 heterodimer dissociated. The bands were quantitated and the fraction of the four-component complex remaining at each time point was determined. The remaining fraction values were plotted versus time, and the data were fit with Equation 17 to obtain a dissociation rate constant (k_{-1}) of 1.8 x 10^{-4} s^{-1}:

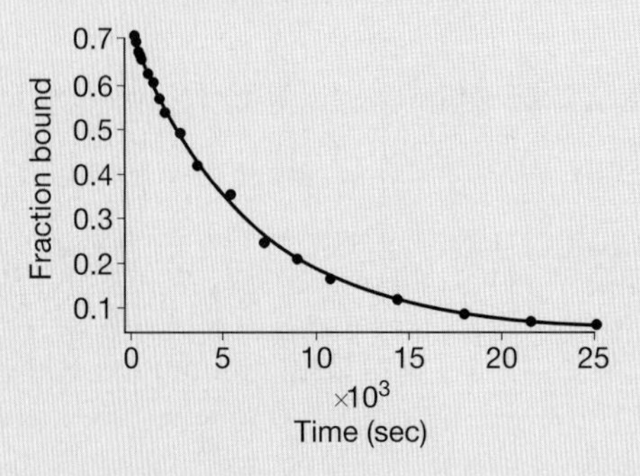

As can be seen from the gel and the plot, the fraction remaining at the first time point (30 seconds) was approximately 0.7. Therefore, it was best to use Equation 17 rather than Equation 13 to fit the data and obtain k_{-1}.

curve is derived from fitting all the data points using Equation 13 (instead of Equation 17), which by definition requires that the curve begin at a fraction remaining of 1. It is obvious that Equation 13 does not generate a curve that adequately fits the data. Moreover, the rate constant measured using Equation 13 is almost twofold greater than the true rate constant for dissociation.

What to do when $[AB]_t/[A]_i$ levels out above 0. When measuring the dissociation of a bimolecular complex, it is possible that the curve will approach a value greater than zero. This would suggest that the dissociation of all AB complexes in a reaction is not governed by a single rate constant. For example, a portion of the A and B associated together might be in the form of aggregates that cannot be measured separately from bimolecular AB complexes. These aggregates might have great kinetic stability and therefore dissociate with a much lower rate constant than the bimolecular AB complexes, leaving the fraction remaining greater than zero at the longest time points. Equation 18 allows k_{-1} to be determined from such data.

$$\frac{[AB]_t}{[AB]_i} = (1 - f_{min})\,(e^{-k_{-1}t}) + f_{min} \tag{18}$$

This equation derives from Equation 13, but it contains the additional variable f_{min}, which defines the point at which the decay curve levels out. When f_{min} is equal to zero, Equation 18 is identical to Equation 13. When the fraction remaining ($[AB]_t/[AB]_i$) is plotted on the *Y* axis and time on the *X* axis, a curve-fitting program can be used to simultaneously solve for k_{-1} and f_{min}.

The solid cure in Figure 4.7 was generated using Equation 18, a k_{-1} value of 0.05 min^{-1}, and an f_{min} value of 0.3. The dashed curve was derived by fitting the data points using Equation 13, which requires that the curve level out at zero. It is obvious that Equation 13 does not generate a curve that adequately fits the data. Moreover, the rate constant measured using Equation 13 is twofold lower than the true rate constant for dissociation of AB complexes.

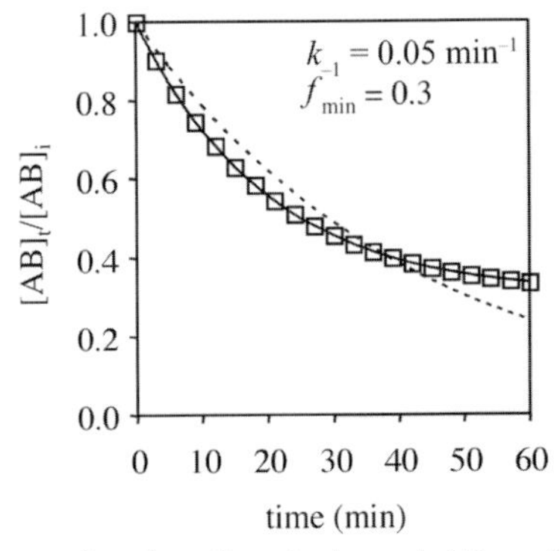

FIGURE 4.7. Decay curve for the dissociation of AB under conditions where a portion of the A associated with B does not dissociate over the time course of the experiment. The k_{-1} and f_{min} used along with Equation 18 to generate the data points and the solid curve are shown. The *dashed curve* demonstrates the poor fit of the data obtained using Equation 13 (the k_{-1} determined using Equation 13 is 0.024 min^{-1}).

General Equations for Fitting Decay Data

The next two equations (Equations 19 and 20) can be used to obtain dissociation rate constants from most bimolecular dissociation data. Both equations take into account (1) the possibility of a rapid decrease in the amount of AB complex upon blocking new association and (2) the possibility that a portion of the AB complex dissociates very slowly or not at all; however, the equations can also be used to fit data sets for which these possibilities do not occur. The two equations differ only with respect to the experimental measurements made. When the fraction remaining is measured ($[AB]_t/[AB]_i$), the following equation should be used to fit the data (for derivation, see Appendix 1):

$$\frac{[AB]_t}{[AB]_i} = (f_{max} - f_{min})(e^{-k_{-1}t}) + f_{min} \qquad (19)$$

Alternatively, when the amount of AB complex is measured (AB_t) the following equation should be used to fit the data:

$$AB_t = (AB_{max} - AB_{min})(e^{-k_{-1}t}) + AB_{min} \qquad (20)$$

The terms f_{max} (Equation 19) and AB_{max} (Equation 20) each define the true starting point of the relevant decay curve (i.e., the Y intercept of the curve at $t = 0$ obtained from fitting the data to one of the two equations). The terms f_{min} (Equation 19) and AB_{min} (Equation 20) each define the point at which the decay curve levels out. Under conditions where f_{max} is 1 and f_{min} is 0, Equation 19 simplifies to Equation 13. Similarly, under conditions where AB_{max} (obtained from fitting the data) equals AB_i (the amount of AB complex measured just before blocking new AB formation) and AB_{min} is 0, Equation 20 simplifies to Equation 15 (see also Simulation S4-1).

To generate a decay curve, either the fraction remaining ($[AB]_t/[AB]_i$) or AB_t is plotted on the Y axis and time on the X axis. A curve-fitting program can be used to simultaneously solve for k_{-1}, f_{max}, and f_{min} in the case of Equation 19, or for k_{-1}, AB_{max}, and AB_{min} in the case of Equation 20 (See Chapter 7). In situations where it is likely that a rapid initial dissociation has occurred, the zero time point is not included in the data set used for the computer fitting.

The plot in Figure 4.8 provides an example of how Equation 20 can be used to fit data that exhibit both a rapid dissociation of a portion of the AB complex and levels out at a point greater than 0. In this example, the AB complex was measured using phosphorimagery, and phosphorimager units were plotted versus time. The data points and curve were generated

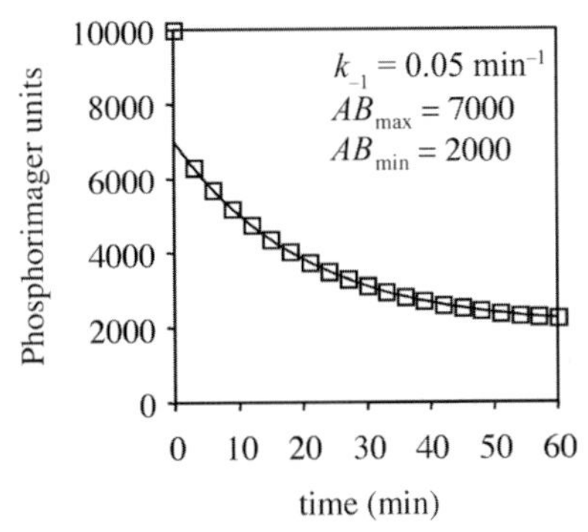

FIGURE 4.8. Decay curve for the dissociation of AB under conditions where a portion of the AB complexes dissociates rapidly, a second portion dissociates with a measurable rate constant, and a third portion does not dissociate over the time course of the experiment. The dissociation of AB was monitored using phosphorimagery (e.g., A was radioactively labeled) and the phosphorimager units were plotted versus time. The k_{-1}, AB_{max}, and AB_{min} used along with Equation 20 to generate the data points and the curve are shown.

using Equation 20, a k_{-1} value of 0.05 min^{-1}, an AB_{max} value of 7000 phosphorimager units, and an AB_{min} value of 2000 phosphorimager units.

Summary: How to experimentally measure k_{-1}

1. Develop an assay to monitor AB.
2. Combine A and B and allow AB complexes to form.
3. Alter conditions to block new AB formation ($t = 0$).
4. Measure the disappearance of AB over time.
5. Plot data and determine k_{-1} (using either Equation 19 or 20).
6. If artifactual decay is suspected, k_{-1} is an upper limit.

ASSOCIATION OF TWO BIOMOLECULES (A + B → AB)

Recall that bimolecular binding interactions can be described by the expression:

$$A + B \underset{k_{-1}}{\overset{k_1}{\rightleftharpoons}} AB \tag{21}$$

The top arrow indicates the association of the two molecules, which is governed by an association rate constant (k_1) with the units of $M^{-1}s^{-1}$. In the following sections, we discuss measurement of association rate constants for the formation of bimolecular biological complexes. Note that we covered much of the theory behind the equations shown below in "General Rate Theory" at the beginning of this chapter.

Theory and Equations

Relating the association rate constant* $\mathbf{k_1}$ *to concentrations of the interacting molecules. The equation that describes the accumulation of AB over time after A and B are combined is:

$$\frac{\text{change in [AB]}}{\text{change in time}} = k_1[\text{A}][\text{B}] - k_{-1}[\text{AB}] \tag{22}$$

where [A], [B], and [AB] are molar concentrations. Determining the association rate constant (k_1) would be most simply done under conditions where the rate of dissociation (k_{-1}[AB]) is far less than the rate of association (k_1[A][B]). In this situation, the dissociation of the AB complex would be negligible, and the measured rate at which AB accumulates would approximate the rate of association of A and B. Therefore, Equation 22 could be simplified to:

$$\frac{\text{change in [AB]}}{\text{change in time}} = k_1[\text{A}][\text{B}] \tag{23}$$

It is typically not possible to block dissociation of the AB complex so that association can be measured in isolation. However, as detailed below, it is not necessary to work under these conditions to measure the association rate constant. In the discussion that follows, we describe a general approach, involving the use of observed rate constants, which can be used to determine k_1 values for single-step bimolecular association reactions.

Relationship between $\mathbf{k_1}$ ***and the observed rate constant for association*** **($\mathbf{k_{obs}}$).** Typically, k_1 is not measured directly. Instead, observed rate constants for association (k_{obs}) are measured at multiple different B concentrations, and k_1 is subsequently determined from these data. The exponential equation to determine an observed rate constant for the accumulation of AB complex as a reaction approaches equilibrium after mixing A and B is (for derivation, see Appendix 1):

$$\boxed{\frac{[\text{AB}]_t}{[AB]_{\text{max}}} = 1 - e^{-k_{\text{obs}}t}} \tag{24}$$

where $[\text{AB}]_t$ is the concentration of AB at each time point and $[AB]_{\text{max}}$ is the concentration of AB at later time points (the plateau value). Here, k_{obs} is a measured rate constant with units of s^{-1}.

When k_{obs} is being measured, it is important to work under conditions where the concentration of free B in the reaction will not effectively change as the AB complex forms (i.e., $[B] \approx [B]_i$). This will always be the case when the initial concentration of B is much greater than the concentration of AB complex at late time points ($[B]_i >> [AB]_{max}$). Typically, this condition is met by setting the initial concentration of B to be much greater than the initial concentration of A (e.g., 100-fold). Under these conditions, the amount of A limits the amount of AB complex that can form, and the concentration of B will not appreciably change as the reaction progresses.

When $[B] \approx [B]_i$, the relationship between k_{obs}, k_1, k_{-1}, and $[B]_i$ is (for derivation, see Appendix 1):

$$k_{obs} = k_1[B]_i + k_{-1} \tag{25}$$

This equation describes a line when the k_{obs} values measured at different concentrations of B are plotted on the *Y* axis and $[B]_i$ values are plotted on the *X* axis. The slope is equal to k_1 and the *Y* intercept is equal to k_{-1}, both of which can be determined by fitting the data using linear regression. The reason that the *Y* intercept is equal to k_{-1} is because as the concentration of B approaches zero, the time that it takes to reach equilibrium will be set by the rate of dissociation.

Another important concept regarding k_{obs} is made clear from Equation 25. The value of k_{obs} will change with the concentration of B; as the concentration of B is increased, the value of k_{obs} increases. This is illustrated in Figure 4.9. The solid curve was generated utilizing a combination of Equations 24 and 25, set values for k_1 and k_{-1}, and a single con-

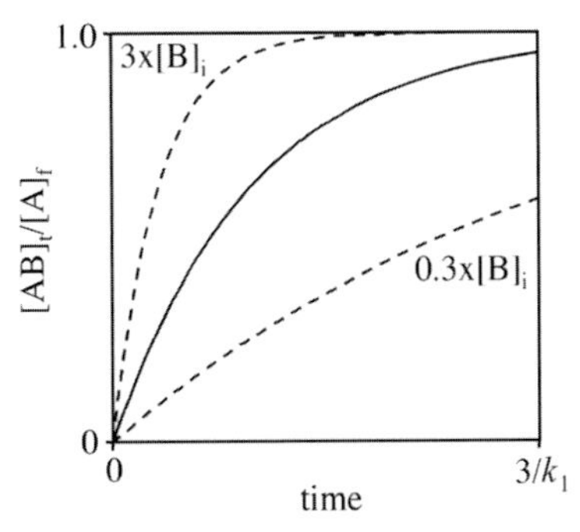

FIGURE 4.9. Dependence of an association rate curve on the concentration of B. The *solid curve* shows the exponential increase in AB over time at a given concentration of B. Shown is the effect of increasing (*left dashed curve*) and decreasing (*right dashed curve*) $[B]_i$ on the rate of formation of the AB complex. A larger k_{obs} is measured with an increase in the concentration of B and a smaller k_{obs} is measured with a decrease in the concentration of B. The progress curves are shown over the range of time from zero to $3/k_1$. Note that k_1 does not change with the concentration of B, and k_{-1} is negligible in the examples shown.

centration of $[B]_i$. When $[B]_i$ was increased and decreased, the k_{obs} increased and decreased, respectively (dashed curves). For example, the dashed curve on the left, which resulted from a threefold increase in $[B]_i$, yields a k_{obs} value that is threefold greater than that obtained from the solid curve. (Recall that the k_{obs} is equal to the inverse of the time at which the reaction is 63% complete). Importantly, the values for k_1 and k_{-1} were kept constant in generating the three curves, and these rate constants do not depend on the concentration of B.

Half-time. In some cases, it may be useful to refer to the half-time for an association reaction. This is the time after mixing A with B at which half of the maximal AB complex has formed. Under the experimental conditions described above ($[B]_i >> [AB]_{max}$ so that the concentration of B does not effectively change with time), the equation for half-time ($t_{1/2}$) is (for derivation, see Appendix 1):

$$t_{1/2} = \frac{\ln 2}{k_{obs}} \tag{26}$$

Because $\ln(2) = 0.693$, the $t_{1/2} = 0.693/k_{obs}$. Importantly, the $t_{1/2}$ for an association reaction is based on k_{obs} and is therefore concentration-dependent (i.e., $t_{1/2}$ will decrease as $[B]_i$ increases).

Experimental Considerations

Developing an assay. Measuring k_{obs} values for association of A with B, then ultimately determining k_1, requires an assay that accurately measures the formation of AB over time. The type of assay depends on the nature of the interaction (e.g., protein–protein or protein–nucleic acid). The assay must distinguish AB complex from unbound A and B such that the concentration of AB complex can be determined at each time point. Often, one biomolecule is labeled (e.g., with radioactivity or a fluorophore) and it does not matter which. Typically, the concentration of one molecule (B in the equations described above) is held well above the concentration of the other molecule (A) to satisfy the condition $[B]_i >> [AB]_{max}$. It is not necessary to have the concentration of B greater than the K_D; however, this might be important in some situations to allow the AB complex to be readily detected. It is important to note that the accuracy with which an association rate constant is measured is only as good as the accuracy with which the concentrations of A and B are known. For a complete discussion of quantitating biomolecules and assessing their fractional activities, see Chapter 1.

Establishing a time course. A tricky aspect of measuring association rate constants is that the time courses appropriate for the measurements often span seconds (or less) as opposed to minutes, making manual sampling impractical. Consider a reaction that has a k_1 of 1×10^6 M^{-1}s^{-1} (a moderate association rate constant for biological binding reactions) and a k_{-1} of 1×10^{-2} s^{-1}. When $[B]_i = 100$ nM, k_{obs} will equal 0.11 s^{-1} (Equation 25) and $t_{1/2}$ will equal 6.3 seconds (Equation 26). Ideally, the time course should begin tenfold below and end tenfold above the $t_{1/2}$. Therefore, in this example, accurately determining the k_{obs} for association would require that a time course be taken from 0.6 to 60 seconds. Clearly, this would be impossible to do via manual sampling.

If manual sampling is the only option, it might be possible to place a lower limit on the association rate constant. For example, if the reaction is complete even at the earliest time point that can be taken manually (e.g. ~5 seconds), it would be fair to state that k_{obs}, and hence k_1, is greater than or equal to 0.2 s^{-1}. In this situation, it might also be useful to estimate k_1 from measurements of K_D and k_{-1} (as discussed in Chapter 2, $k_1 = k_{-1}/K_D$).

Determining k_{obs} by measuring AB. Often, it is most convenient to simply quantitate the amount of AB complex formed over time as opposed to determining the concentration of AB complex. When AB is measured, the following equation should be used to determine the k_{obs} (for derivation, see Appendix 1):

$$AB_t = AB_{max}(1 - e^{-k_{obs}t}) \qquad (27)$$

Here, AB_t values are not concentrations, but instead are the amounts of AB complex present at the different time points, quantitated in a manner specific to the assay. For example, AB_t could be expressed in phosphorimager units, fluorescence emitted, scintillation counts, etc. AB_{max} is the maximum amount of AB complex that can form at long times; it will have the same units as AB_t.

Consider the example shown in Figure 4.10. Here, B was ^{32}P-labeled and the AB complex was isolated and quantitated by scintillation counting. Scintillation counts were plotted on the Y axis and time on the X axis. A and B were mixed at $t = 0$. As time increases, the scintillation counts in the AB complex increase to a plateau defined by AB_{max}. Both k_{obs} and AB_{max} can be obtained by fitting the data with a nonlinear regression program using Equation 27 (see Chapter 7). The half-time (i.e., the time at which 50% of the A in the reaction is in AB complexes) is 1.7 seconds ($t_{1/2} = 0.693/0.4$).

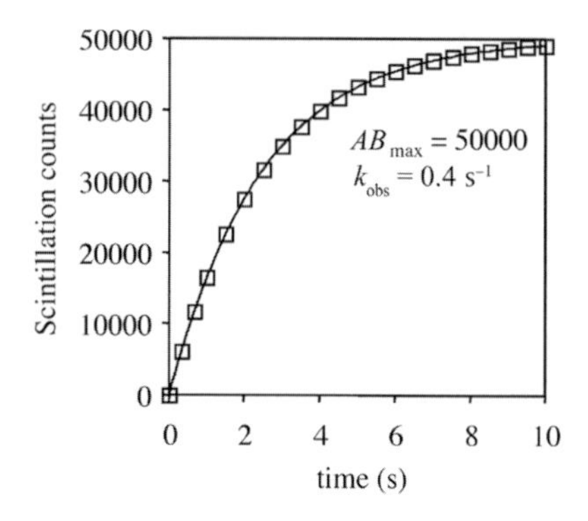

FIGURE 4.10. Association curve for the formation of AB, obtained by plotting scintillation counts (i.e., the amount of AB) versus time. The AB_{max} and k_{obs} used with Equation 27 to generate the data points and curve are shown.

Assessing the relationship between $[B]_i$ and k_{obs}. According to Equation 25, there is a linear relationship between k_{obs} and $[B]_i$. k_1 and k_{-1} can be determined by measuring k_{obs} at multiple concentrations of B, plotting k_{obs} and $[B]_i$ on the *Y* axis and *X* axis, respectively, and using linear regression to fit the data (see Chapter 7). It is best to measure k_{obs} values at four or more concentrations of B that span at least a tenfold range. An example plot is shown in Figure 4.11 (see also Simulation S4-2). Here, k_{obs} values were measured at five concentrations of B, from 2 to 40 nM. The slope of the plot, and hence k_1, is equal to 1×10^7 M–1s^{-1}. The *Y* intercept of the plot is equal to 0.01 s^{-1}. It is important to realize that the *Y* intercept is

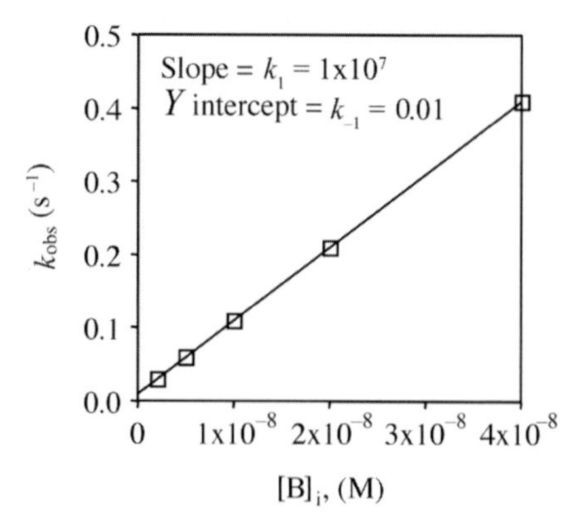

FIGURE 4.11. Plot of k_{obs} values for association versus $[B]_i$ for a single-step binding reaction. The k_1 is equal to 1×10^7 $M^{-1}s^{-1}$ (the slope of the line). The k_{-1} is equal to 0.01 s^{-1} (the *Y* intercept of the line).

often too close to zero to allow k_{-1} to be accurately determined using this approach. Hence, it is usually best to measure k_{-1} directly using the methods described in the previous section of this chapter.

Diffusion limited. The fastest rate at which two molecules can interact is ultimately limited by the rate of diffusion; biomolecules must move through solution to interact with one another. Moreover, not all contacts between two interacting molecules result in functional interactions (e.g., the interacting surfaces must be in the correct orientation to one another). If a measured k_1 is on the order of 10^8 $\text{M}^{-1}\text{s}^{-1}$, it is likely that the rate of association is diffusion limited. Surprisingly, the k_1 values for some biomolecular interactions are greater than 10^8 $\text{M}^{-1}\text{s}^{-1}$. This can be explained via a multistep mechanism for binding, where steps after the first contact occur in fewer than three dimensions and hence are not limited by the rate of three-dimensional diffusion. For example, a protein that binds DNA at a specific site might first interact with the DNA at some random site and then scan the DNA (via one-dimensional diffusion) until the specific site is located. When measuring association rate constants, it is useful to keep in mind the limitation imposed by three-dimensional diffusion. If a k_1 significantly greater than 10^8 $\text{M}^{-1}\text{s}^{-1}$ is measured, it is likely that the binding reaction being studied has more than one step, which could prove very interesting and biologically important. We consider two-step bimolecular reactions in detail in Chapter 5.

Summary: How to experimentally measure k_1

1. Develop an assay to monitor AB.
2. Combine A and B at $t = 0$ ($[\text{B}]_i >> [AB]_{max}$).
3. Measure accumulation of AB over time.
4. Plot data and determine k_{obs} (using Equation 24 or 27).
5. Repeat Steps 2–4 at multiple different $[\text{B}]_i$.
6. Plot k_{obs} versus $[\text{B}]_i$ and determine k_1 (Equation 25).
7. Consider whether k_1 is diffusion limited.

Literature Example 4.3. Determining the association rate constant for a protein–DNA complex. *(Jamieson E.R. and Lippard S.J. 2000. Biochemistry 39: 8426–8438.) Figures reprinted, with permission, from Jamieson and Lippard (2000). (©American Chemical Society)*

The anticancer drug cisplatin is used to treat a variety of human tumors. It binds to DNA and forms intrastrand cross-links, or adducts, that distort the structure of the DNA and thereby inhibit replication and transcription. These adducts are recognized by cellular proteins, including the high-mobility group (HMG) domain protein family. This study characterized the kinetic parameters that govern the formation of protein–DNA complexes between HMG1 domains A and B and platinated DNA probes (i.e., containing a cisplatin DNA adduct).

The authors used stopped-flow fluorescence spectroscopy to assay the formation of HMG1 domain-platinated DNA complexes over time. A fluorescein-modified deoxyuridine residue was present in the major groove of several platinated 16-bp DNA probes. When the HMG1 domain protein bound to the DNA probe in the minor groove, a large increase in fluorescence intensity was observed.

Observed rate constants for association were measured using six different HMG1 domain A concentrations ranging from 250 to 1200 nM, with the concentration of the DNA probe held constant at 25 nM. Binding was initiated by rapid mixing of a DNA probe with HMG1 domain A. Data points were collected over time from 0 to 0.15 second. The data were plotted with fluorescence on the *Y* axis and time (s) on the *X* axis for each concentration of HMG1 domain A, as shown in the plot below.

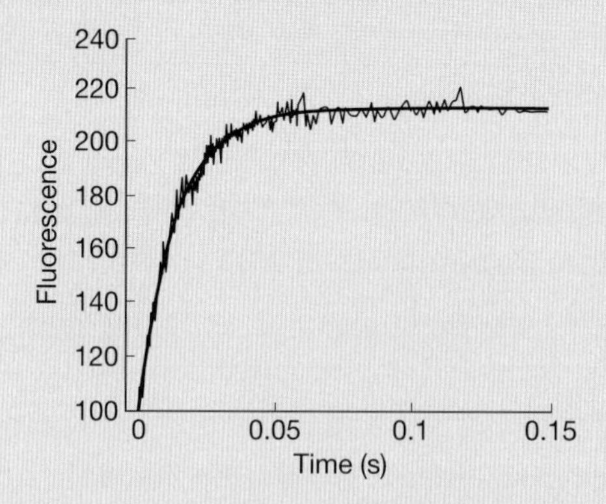

Observed rate constants were determined using Equation 27 and subsequently plotted versus the concentrations of HMG1 domain A. The points were fit with a line, and the slope, corresponding to the association rate constant (k_1), was determined.

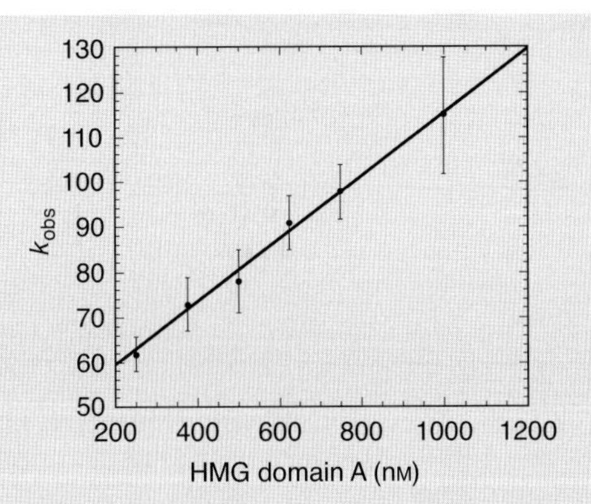

Overall, the authors concluded that binding of HMG1 domains A and B to multiple different platinated DNA probes is fast, with k_1 values approaching the diffusion limit (~10^8 $M^{-1}s^{-1}$).

The dissociation rate constants (k_{-1}), determined from the *Y* intercepts of the plots of k_{obs} versus protein concentration, ranged from about 30 to 150 s^{-1}. Often, this not the most accurate way to determine k_{-1}; therefore, the authors performed additional experiments in which the dissociation rate constants were directly measured. In this case, the values for k_{-1} using the two different approaches agreed within error.

CHAPTER 5

Kinetics of Regulation and Cooperativity in Binding Reactions

IN THIS CHAPTER, WE CONSIDER THE KINETICS that govern the formation of three component complexes, in terms of both regulators of bimolecular interactions and cooperativity. Many trimolecular complexes are found in biology; for example, two different proteins that bind to adjacent sites on a molecule of DNA. Importantly, regulators of bimolecular binding reactions can often be modeled as three component interactions. For example, if a small molecule binds a protein and increases its affinity for binding a target protein, the small molecule is a regulator of the protein–protein interaction. This situation can be modeled as a three-component system that exhibits positive cooperativity and can be studied using the approaches we describe here. Throughout the chapter, hypothetical illustrations and examples from the literature are provided to demonstrate practical aspects of measuring rate constants. Key equations are contained in boxes and are derived in Appendix 1. In addition, a computer simulation is provided on the Web site (http://kinetics.cshl.edu) that can be manipulated to help visualize concepts presented in the chapter.)

> ***Topics covered:***
>
> - Three-component interactions (A + B + C $\leftrightarrow$ ABC)
> - Rate constants can reveal mechanisms of regulation and cooperativity
> - Measuring association and dissociation rate constants in three-component systems

TERMS AND PRINCIPLES

Rate of a binding reaction describes the change in the concentration of a molecule or complex per change in time. The units for the rate of a binding reaction are $\text{M}t^{-1}$ (M/t), where M is molar and t is time (e.g., seconds or minutes).

Rate constants for binding reactions are numeric representations of the time that it takes for molecules to associate or dissociate. In general, rate = (rate constant)(concentrations).

Association rate constant **(k_1)** relates the rate at which biomolecules associate to the concentrations of the molecules. The rate constant k_1 has units of $\text{M}^{-1}t^{-1}$, where M is molar and t is time (e.g., seconds or minutes). In general, the larger the value of k_1, the faster the inherent rate of association between A and B.

Dissociation rate constant **(k_{-1})** relates the rate at which biomolecules in a complex dissociate to the concentration of the complex. The rate constant k_{-1} has units of t^{-1}, where t is time (e.g., seconds or minutes). In general, the larger the value of k_{-1}, the faster the inherent rate of dissociation.

Cooperativity occurs in systems with three or more interacting molecules when the binding of one molecule affects the binding of others, either positively or negatively. For example, in the three-component interaction A + B + C ↔ ABC, cooperativity would exist if the binding of C to A affected the affinity of A for B. Cooperativity can be either positive or negative. Positive cooperativity exists when the presence of C increases the affinity of A for B (in this case, C can be thought of as an activator of the AB interaction). Negative cooperativity exits when the presence of C decreases the affinity of A for B (in this case, C can be thought of as a repressor of the AB interaction).

The cooperativity factor α describes the level of cooperativity that arises when one biomolecule affects the affinity with which two other biomolecules interact. Values of α are always greater than zero. When α is less than 1, a system has positive cooperativity. When α is greater than 1, a system has negative cooperativity. When α equals 1, a system lacks cooperativity.

Observed rate constants **(k_{obs})**, as explained in this chapter, are experimentally measured values that describe the formation of a complex over time. Observed rate constants mathematically relate to the actual rate constants for association (k_1) and dissociation (k_{-1}), and have the units of t^{-1}, where t is time (e.g., seconds or minutes).

KINETICS OF THREE-COMPONENT INTERACTIONS (A + B + C ↔ ABC)

We focus here on understanding the kinetics for the formation of trimolecular complexes by measuring the rate constants for association and dissociation that govern the bimolecular complexes that form en route to the higher-order complex. This understanding can provide insight into the mechanisms underlying regulation or cooperativity in three-component systems. Much of the following discussion draws on information presented in Chapter 3 (Cooperativity in Binding) and Chapter 4 (Rate Constants for Binding and Dissociation). It may therefore be helpful to refer back to these chapters for further explanation.

Theory and Equations

Modeling trimolecular interactions. Trimolecular binding interactions can be described by the following model, which considers all possible bimolecular complexes that can form en route to ABC (AB, AC, and BC):

$$\begin{array}{ccc} \text{A} + \text{B} & \rightleftharpoons & \text{AB} \\ + & \searrow\!\!\nwarrow & + \\ \text{C} & \text{BC} + \text{A} & \text{C} \\ \upharpoonleft\!\downharpoonright & \searrow\!\!\nwarrow & \upharpoonleft\!\downharpoonright \\ \text{AC} + \text{B} & \rightleftharpoons & \text{ABC} \end{array} \tag{1}$$

Contacts between all three molecules could contribute to the formation and stability of the ABC complex (see Fig. 3.1 in Chapter 3). In many cases, however, one of the three bimolecular complexes forms quite weakly at biological concentrations. For example, two proteins (B and C) that bind neighboring sites on DNA cooperatively might make weak protein–protein contacts. These contacts are important for cooperativity between the DNA-bound proteins, but they are too weak to allow interaction between the proteins at physiological concentrations. Therefore, experiments can be performed under conditions where the protein–protein complex (BC) does not form. In these cases, Equation 1 can be simplified to Equation 2, in which the BC complex is not considered.

$$\begin{array}{ccc} \text{A} + \text{B} & \underset{k_{-1}}{\overset{k_1}{\rightleftharpoons}} & \text{AB} \\ + & & + \\ \text{C} & & \text{C} \\ k_{-2}\Big\uparrow\!\Big\downarrow k_2 & & k'_{-2}\Big\uparrow\!\Big\downarrow k'_2 \\ \text{AC} + \text{B} & \underset{k'_{-1}}{\overset{k'_1}{\rightleftharpoons}} & \text{ABC} \end{array} \tag{2}$$

This model is easier to understand and study because it only considers two of the bimolecular complexes that could possibly form (AB and AC). Here, the arrows pointing to the right or down indicate associations, which are governed by association rate constants (k_1, k_2, k'_1, k'_2) with units of $\text{M}^{-1}\text{s}^{-1}$. The arrows pointing to the left or up indicate dissociations, which are governed by dissociation rate constants (k_{-1}, k_{-2}, k'_{-1}, k'_{-2}) with units of s^{-1}. A prime (′) is used to denote rate constants in the presence of a third molecule. For example, k'_1 is the rate constant for association of A and B in the presence of C, whereas k_1 is the rate constant for association of A and B in the absence of C. In this example, molecule C is a regulator of the interaction between A and B. The model in Equation 2 can also be depicted by the combination of two linear equations:

$$\begin{gathered} A + B \underset{k_{-1}}{\overset{k_1}{\rightleftharpoons}} AB + C \underset{k'_{-2}}{\overset{k'_2}{\rightleftharpoons}} ABC \\ A + C \underset{k_{-2}}{\overset{k_2}{\rightleftharpoons}} AC + B \underset{k'_{-1}}{\overset{k'_1}{\rightleftharpoons}} ABC \end{gathered} \tag{3}$$

In the following sections, we describe the theory, equations, and experimental principles used to evaluate regulation, cooperativity, and rate constants in systems that fit the model shown in Equations 2 and 3. We primarily use examples in which C is a regulator of the binding reaction between A and B.

Using kinetics to study regulators of binding reactions and mechanisms of cooperativity. In systems with three interacting molecules, regulation or cooperativity occurs when the binding between two biomolecules affects the binding of the third, either positively or negatively. As first described in Chapter 3 (see Fig. 3.3), there are, at the molecular level, two common mechanisms that give rise to regulation or cooperativity in trimolecular interactions: (1) weak contacts between B and C affecting the overall formation of the ABC complex and (2) the binding of C to A, inducing a conformational change in A that alters its affinity for B.

Kinetics can be used to better understand mechanisms of regulation and cooperativity by determining how C affects the rate constants governing the association and dissociation of A and B. Consider again a system with positive cooperativity, where the cooperativity is due to weak contacts between two proteins (B and C) that bind DNA. These contacts might allow B to associate more rapidly with an AC complex than with A alone. Alternatively, the presence of C could help hold B in a trimolecular complex, thereby decreasing the rate of dissociation of A and B. In both of these cases, C enhances the association between A and B, resulting in positive cooperativity (therefore, C would be considered an activator of the binding reaction between A and B). More complex models are also possible, in which the rates of both association and dissociation of A with B are affected by C. Kinetics can also be used to describe mechanisms of regulation and cooperativity resulting from a conformational change as opposed to weak interactions between B and C.

In the remainder of the chapter, we primarily use the term cooperativity (as opposed to regulation) when describing how C can affect the binding of A with B. It is important to understand, however, that the mechanisms governing cooperative binding apply to systems in which the regulator of a bimolecular interaction is being studied (we illustrate this point in Literature Example 5.1).

Rate Constants Can Reveal Mechanisms of Regulation and Cooperativity

Relating Rate Constants to the Cooperativity Factor α. The level of cooperativity in a system is defined by the cooperativity factor α, which can be measured by comparing the K_D for a bimolecular interaction in the absence and presence of a third molecule, as summarized below. Recall that a K_D = (dissociation rate constant)/(association rate constant) (e.g., k_{-1}/k_1).

Summary of cooperativity, binding affinity, and the cooperativity factor α *(adapted from Chapter 3)*

Consider the following equations that describe the formation of ABC when both the AB and AC complexes can form, but BC cannot:

$$\begin{array}{ccc} & K_{D(AB)} & \\ A + B & \rightleftharpoons & AB \\ + & & + \\ C & & C \\ K_{D(AC)} \updownarrow & & \updownarrow \alpha K_{D(AC)} \\ AC + B & \rightleftharpoons & ABC \\ & \alpha K_{D(AB)} & \end{array} \tag{4}$$

This model can also be depicted by the combination of two linear equations:

$$\begin{array}{l} A + B \overset{K_{D(AB)}}{\rightleftharpoons} AB + C \overset{\alpha K_{D(AC)}}{\rightleftharpoons} ABC \\ \\ A + C \overset{K_{D(AC)}}{\rightleftharpoons} AC + B \overset{\alpha K_{D(AB)}}{\rightleftharpoons} ABC \end{array} \tag{5}$$

$K_{D(AB)}$ is the equilibrium dissociation constant for B binding A, and $K_{D(AC)}$ is the equilibrium dissociation constant for C binding A.

(Continued on following page)

Summary of cooperativity, binding affinity, and the cooperativity factor α *(Continued)*

The term α describes the cooperativity in the system. For example, the effect of C on the K_D for the interaction of A with B is represented by α, in that $K_{D(AB)}$ becomes $\alpha K_{D(AB)}$ in the presence of C. If there is positive cooperativity, α is less than 1; the presence of C increases the affinity of A for B (i.e., $\alpha K_{D(AB)} < K_{D(AB)}$). If there is negative cooperativity, α is greater than 1; the presence of C decreases the affinity of A for B (i.e., $\alpha K_{D(AB)} > K_{D(AB)}$). If there is no cooperativity in the system, α is equal to 1 and $\alpha K_{D(AB)}$ is equal to $K_{D(AB)}$. These concepts are summarized in Table 5.1.

It is important to understand that although Equations 4 and 5 contain α twice, it is a single constant with one value for a given system. In other words, if C affects the affinity of A for B by a factor of α, then B affects the affinity of A for C by the same factor of α (i.e., $\alpha K_{D(AB)}$ and $\alpha K_{D(AC)}$ can have different values, but α must be the same).

The following equations relate rate constants to affinity constants for the models shown in Equations 2–5.

$$K_{D(AB)} = \frac{k_{-1}}{k_1} \tag{6}$$

$$K_{D(AC)} = \frac{k_{-2}}{k_2} \tag{7}$$

$$\alpha K_{D(AB)} = \frac{k'_{-1}}{k'_1} \tag{8}$$

$$\alpha K_{D(AC)} = \frac{k'_{-2}}{k'_2} \tag{9}$$

Substituting Equation 6 into Equation 8 gives the following equation, which relates α to the rate constants for association and dissociation of A

TABLE 5.1. How Cooperativity Relates to Binding Affinity

Value for α	Cooperativity	Effect on affinity	Effect on K_D
$\alpha < 1$	positive	increase	decrease
$\alpha = 1$	none	none	none
$\alpha > 1$	negative	decrease	increase

with B, in the presence and absence of C:

$$\alpha \frac{k_{-1}}{k_1} = \frac{k'_{-1}}{k'_1} \tag{10}$$

Similarly, substituting Equation 7 into Equation 9 gives the following equation, which relates α to the rate constants for association and dissociation of A with C, in the presence and absence of B:

$$\alpha \frac{k_{-2}}{k_2} = \frac{k'_{-2}}{k'_2} \tag{11}$$

Importantly, the overall equilibrium dissociation constant describing the formation of ABC is independent of the order in which components assemble. Therefore, $K_{D(AB)}\alpha K_{D(AC)} = K_{D(AC)}\alpha K_{D(AB)}$ (for further explanation, see Chapter 3). This relationship can be expressed in terms of rate constants by combining Equations 10 and 11 to yield:

$$\frac{k_{-1}k'_{-2}}{k_1 k'_2} = \frac{k_{-2}k'_{-1}}{k_2 k'_1} \tag{12}$$

Useful generalizations can be made concerning the relationship between α and these eight rate constants. For example, if there is positive cooperativity in the system (i.e., $\alpha < 1$), then $k'_{-1}/k'_1 < k_{-1}/k_1$ (see Equation 10). If there is negative cooperativity in the system (i.e., $\alpha > 1$), then $k'_{-1}/k'_1 > k_{-1}/k_1$. If there is no cooperativity in the system (i.e., $\alpha = 1$), then $k'_{-1}/k'_1 = k_{-1}/k_1$. Also consider that rearranging Equation 10 gives the following equation:

$$\frac{k'_{-1}}{k_{-1}} = \alpha \frac{k'_1}{k_1} \tag{13}$$

Here, the ratio between a pair of rate constants measured with and without C (e.g., k'_{-1} and k_{-1}) can be compared to α to understand the mechanism of cooperativity. For example, if $\alpha = 0.1$ and $k'_1/k_{-1} = 0.1$, then k'_1/k_1 must equal 1. Therefore, the rate constant for association of A with B cannot be affected by molecule C because k'_1 must equal k_1 for $k'_1/k_1 = 1$. As a result, in this example, all of the cooperativity results from the effect on the rate constant for dissociation, with the ABC complex being tenfold more stable than the AB complex (e.g., $k'_{-1} = 0.1\ k_{-1}$).

Using kinetic experiments to decipher mechanisms of regulation and cooperativity (as opposed to levels of cooperativity). Determining the overall magnitude of cooperativity in a trimolecular system requires either measuring all eight rate constants shown in Equation 2 or measuring the cooperativity factor α, as described in Chapter 3. In most systems, it is easier to assess the overall level of cooperativity by measuring α directly; however, the value of α reveals nothing about the mechanism by which the cooperativity occurs. Rate constants must be measured to gain insight into the mechanism. In other words, knowing α will not reveal how molecule C affects the association of A with B. Does molecule C increase the rate at which A associates with B? Is the ABC complex more or less kinetically stable than the AB complex? Kinetic studies can be used to answer these questions and therefore provide an understanding of the underlying mechanism of cooperativity in a system. For example, positive cooperativity could result from molecule C causing a decrease in the dissociation rate constant ($k'_{-1} < k_{-1}$), an increase in the association rate constant ($k'_1 > k_1$), or changes in both the dissociation and association rate constants.

Table 5.2 considers some possible effects of C on the interaction between A and B. The values in Table 5.2 depict systems having positive cooperativity (where C is an activator), negative cooperativity (where C is a repressor), or no cooperativity. The two relevant bimolecular reactions and their rate constants are:

$$\mathrm{A} + \mathrm{B} \underset{k_{-1}}{\overset{k_1}{\rightleftharpoons}} \mathrm{AB} \tag{14}$$

$$\mathrm{AC} + \mathrm{B} \underset{k'_{-1}}{\overset{k'_1}{\rightleftharpoons}} \mathrm{ABC} \tag{15}$$

The first row of Table 5.2 lists rate constants for association and dissociation of A and B in the absence of C. The second row represents a noncooperative ($\alpha = 1$) trimolecular binding reaction in which the association and dissociation rate constants for the binding of A to B are unaffected by the presence of C (i.e., $k'_1 = k_1$ and $k'_{-1} = k_{-1}$). The remaining rows exhibit different effects that C could have on the association and dissociation rate constants

TABLE 5.2. Examples of Different Combinations of Rate Constants and α Values

Row	α	k_1 ($\mathrm{M}^{-1}\mathrm{s}^{-1}$)	k'_1 ($\mathrm{M}^{-1}\mathrm{s}^{-1}$)	k_{-1} (s^{-1})	k'_{-1} (s^{-1})
1		1×10^7		1×10^{-2}	
2	1	1×10^7	1×10^7	1×10^{-2}	1×10^{-2}
3	0.2	1×10^7	**5×10^7**	1×10^{-2}	1×10^{-2}
4	0.2	1×10^7	1×10^7	1×10^{-2}	**0.2×10^{-2}**
5	5	1×10^7	**0.2×10^7**	1×10^{-2}	1×10^{-2}
6	5	1×10^7	1×10^7	1×10^{-2}	**5×10^{-2}**
7	0.2	1×10^7	**10×10^7**	1×10^{-2}	**2×10^{-2}**
8	1	1×10^7	**0.3×10^7**	1×10^{-2}	**0.3×10^{-2}**

for the interaction between A and B, as reflected by changes in k_1' and k_{-1}' (highlighted in bold).

Rows 3–6 of Table 5.2 demonstrate simple models of how C (a regulator) might affect rate constants for the interaction between A and B. Rows 3 and 4 represent positive cooperativity ($\alpha < 1$) and rows 5 and 6 represent negative cooperativity ($\alpha > 1$). In general, positive cooperativity (or activation by molecule C) favors the forward reaction by either increasing the rate constant for association or decreasing the rate constant for dissociation (compare rows 3 and 4, respectively, with row 2). In contrast, negative cooperativity (or repression by molecule C) favors the reverse reaction by either decreasing the rate constant for association or increasing the rate constant for dissociation (compare rows 5 and 6, respectively, to row 2).

Rows 7 and 8 demonstrate more complex effects that molecule C could have on the rate constants for association and dissociation of A and B. Row 7 demonstrates how positive cooperativity can result when C increases both k_1 and k_{-1}. Finally, in rare cases, it is possible that the rate constants for association and dissociation both change in the presence of C, but the system is noncooperative (compare row 8 to row 2).

These examples underscore the point that although α describes the level of cooperativity in a system, it does not provide insight into the mechanism of cooperativity. Kinetic experiments can reveal how cooperativity arises (i.e., how a regulator controls a binding reaction).

Experimental Considerations

A total of eight rate constants (all shown in Equation 2) must be measured to assess all possible effects of cooperativity on rates of association and dissociation in a trimolecular system (assuming the BC complex does not form). Although this is not often practical, much can be learned from assessing the effect of one molecule (e.g., the regulator) on the rate constants for association and dissociation of the other two molecules. For example, to determine the effect of C on the rate constant for association of A and B, only two association rate constants need to be measured: (1) k_1 for the association of A and B in the absence of C and (2) k_1' for the association of A and B in the presence of C. Similarly, the effect of C on the dissociation of AB can be determined by measuring (1) k_{-1} for the dissociation of A and B in the absence of C and 2) k_{-1}' for the dissociation of A and B in the presence of C.

In the following sections, we discuss how to measure k_1, k_1', k_{-1}, and k_{-1}' to determine the effect of C on the rate constants for association and dissociation of A and B. The expressions for the two relevant bimolecular interactions are shown in Equations 14 and 15. These same methods can be used to measure k_2, k_2', k_{-2}, and k_{-2}'. The theory and equations presented in Chapter 4 for measuring association and dissociation rate constants in bimolecular systems apply here also. We reiterate below the key points from

Chapter 4, but it may be helpful to refer back to this chapter for a more detailed description of the theory and derivations behind the equations.

***Measuring* k_1 *and* k_1'.** The information in the box below summarizes the approach for measuring k_1, and k_1', and the subsequent discussion expands on these points.

Measuring k_1 and k_1'

1a. Develop an assay to monitor AB.
b. Combine A and B at $t = 0$ ($[B]_i >> [AB]_{max}$).
c. Measure accumulation of AB over time.
d. Plot data and determine k_{obs} (using Equation 16).
e. Repeat Steps b–d at multiple different $[B]_i$.
f. Plot k_{obs} versus $[B]_i$ and determine k_1 (using Equation 18).
2a. Develop an assay to monitor ABC.
b. Combine AC and B at $t = 0$ ($[B]_i >> [ABC]_{max}$).
c. Measure accumulation of ABC over time.
d. Plot data and determine k'_{obs} (using Equation 17).
e. Repeat Steps b–d at multiple different $[B]_i$.
f. Plot k'_{obs} versus $[B]_i$ and determine k_1' (using Equation 19).
3. Compare k_1' to k_1.

1. *Measuring* k_{obs} *and* k'_{obs} *values.* To obtain k_1, k_{obs} values must be measured at multiple concentrations of B. k_{obs} is a measured rate constant for the formation of AB over time with units of s^{-1}. For a reaction performed at a single concentration of B, the following equation describes the exponential accumulation of AB complex as the reaction approaches equilibrium after mixing A with B:

$$AB_t = AB_{max}\,(1 - e^{-k_{obs}t}) \tag{16}$$

Here, AB_t is the amount of AB complex present at each different time point, quantitated in a manner specific to the assay. For example, AB_t could be expressed in phosphorimager units, fluorescence emitted, scintillation counts, etc. AB_{max} is the maximum amount of AB complex that can form at long times; it will have the same units as AB_t. This equation describes a curve that begins at zero and exponentially approaches AB_{max}. When measuring k_{obs}, it is important to work under conditions where the concentration of B ($[B]_i$) is much higher than the final concentration of AB complex that forms at long time points ($[AB]_{max}$). Under these conditions, the concentration of free B in the reaction will not effectively change as the AB complex forms (i.e., $[B] \approx [B]_i$). The simplest way to ensure that this is the case is to set $[B]_i >> [A]_i$.

To obtain k'_1, k'_{obs} values are measured at multiple concentrations of B using an approach similar to that described above, with one key exception. In this case, molecule C is included in reactions at a concentration that allows it to saturate all of the A in the reaction. Unbound C in reactions is not a problem as long as it is not present at concentrations high enough to form the BC complex. This situation therefore mimics a bimolecular system in which the AC complex acts as a single entity (see Equation 15); therefore, the effect of C on the observed rate constant for association of B with AC (k'_{obs}) can be determined. An assay must be developed that distinguishes ABC from AC. This assay is used to measure the amount of ABC complex formed over time and k'_{obs} is determined using the following equation:

$$\boxed{\mathrm{ABC_t} = ABC_{max}\,(1 - e^{-k'_{obs}t})} \qquad (17)$$

There the terms are analogous to those described for Equation 16.

The data and curves in Figure 5.1 were generated using Equations 16 and 17 and together illustrate a positive effect of molecule C on the rate of association of A and B. In this example, A was ^{32}P-labeled and used in two experiments. In the first experiment, A and B were mixed at t = 0, and the AB complex was quantitated by phosphorimagery over time. In the second experiment, the AC complex was formed, mixed with B at t = 0, and the ABC complex was quantitated by phosphorimagery over time. Phosphorimager units for both experiments were plotted on the Y axis and time on the X axis. With increasing time, the phosphorimager units in both the AB (circles) and the ABC (squares) complex-

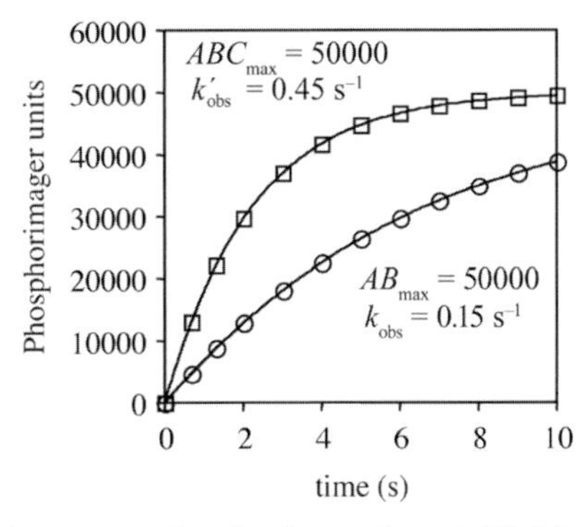

FIGURE 5.1. Association curves for the formation of AB (*circles*) and ABC (*squares*), obtained by plotting phosphorimager units versus time. The parameters used, along with Equations 16 and 17, to generate the data points and curves are shown.

es increase to plateaus defined by AB_{max} and ABC_{max}, respectively. Each data set can be fit individually with a nonlinear regression program (see Chapter 7) using Equation 16 to obtain k_{obs} and AB_{max} in the absence of C and using Equation 17 to obtain k'_{obs} and ABC_{max} in the presence of C. At this concentration of B_i, molecule C causes a threefold increase in the observed rate constant for association (from 0.15 to 0.45 s^{-1}).

2. *Obtaining* k_1 *and* k'_1. The two experimental schemes described above are now repeated, each at multiple concentrations of B, to measure the values of k_{obs} and k'_{obs} for each $[B]_i$. The resulting data sets can be used along with the following equations to obtain k_1 and k'_1:

$$k_{obs} = k_1[B]_i + k_{-1} \quad (18)$$

$$k'_{obs} = k'_1[B]_i + k'_{-1} \quad (19)$$

These equations describe lines with slopes equal to k_1 or k'_1. The value for k_1 can be determined by plotting k_{obs} on the *Y* axis and $[B]_i$ on the *X* axis and using linear regression to fit the data (see Chapter 7). The value for k'_1 can be similarly determined from a plot of k'_{obs} values versus $[B]_i$. The *Y* intercept of each of these plots provides a measurement of k_{-1} or k'_{-1}; however, these values are not accurate unless they are significantly greater than zero. In the following section, we describe a more direct approach to measuring dissociation rate constants (k_{-1} and k'_{-1}).

Example plots illustrating how to determine k_1 and k'_1 are shown in Figure 5.2 (see also Simulation S5). The data shown as circles and squares represent reactions in the absence and presence of molecule C, respectively. The k_1 and k'_1 values (from the slopes of the lines) are 1.2 x

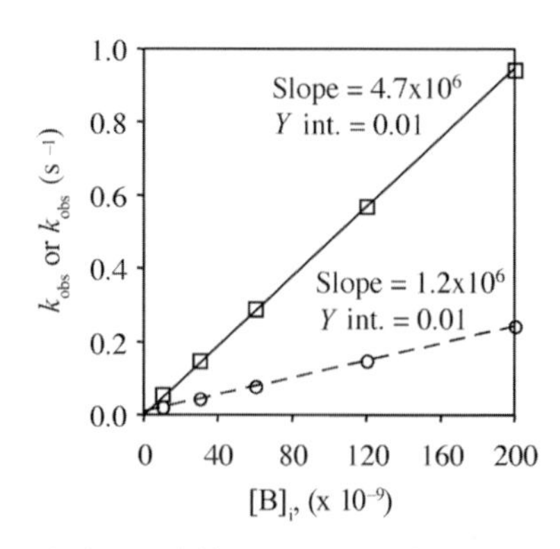

FIGURE 5.2. Plot of k_{obs} (*circles*) and k'_{obs} (*squares*) values for association versus $[B]_i$. The two data sets were fit using linear regression. The slopes and *Y* intercepts of the two lines are shown.

Illustration 5.1. Using kinetics to assess the mechanism of cooperative interactions between transcription factors and promoter DNA.

Transcriptional activator proteins X and Y cooperatively bind to composite elements that are located in the promoters of several tightly regulated genes. You are interested in studying how gene Q, which contains a composite binding site for X and Y, is transcriptionally regulated. Your goal is to determine the mechanism underlying the cooperative binding of X and Y to DNA. As a first step, you decide to perform kinetic assays to measure the association rate constants that drive the formation of the ternary complex between the DNA, protein X, and protein Y.

You write the equations that describe the formation of the ternary complex (analogous to Equation 3):

$$\mathrm{Q} + \mathrm{X} \underset{k_{-1}}{\overset{k_1}{\rightleftharpoons}} \mathrm{QX} + \mathrm{Y} \underset{k'_{-2}}{\overset{k'_2}{\rightleftharpoons}} \mathrm{QXY}$$

$$\mathrm{Q} + \mathrm{Y} \underset{k_{-2}}{\overset{k_2}{\rightleftharpoons}} \mathrm{QY} + \mathrm{X} \underset{k'_{-1}}{\overset{k'_1}{\rightleftharpoons}} \mathrm{QXY}$$

where Q is the DNA containing the composite element to which proteins X and Y bind. You plan on independently measuring k_1 and k'_1. First, you develop an assay to monitor the formation of protein–DNA complexes that enables very rapid time points to be taken. You decide to use surface plasmon resonance in which a double-stranded oligonucleotide containing the composite element from gene Q that binds X and Y is immobilized on a sensor chip via a biotin-streptavadin linkage. The proteins are flowed over the chip and you monitor changes in the resonance units of the sample over time as proteins bind.

To ultimately determine k_1, you first measure the observed rate constants for the association of protein X with Q, using the following concentrations of X (nM): 5, 10, 25, 50, and 100. For each data set, you plot resonance units on the *Y* axis and time on the *X* axis and fit the data with Equation 16 to obtain the curves shown. The concentrations of protein X are indicated next to each curve.

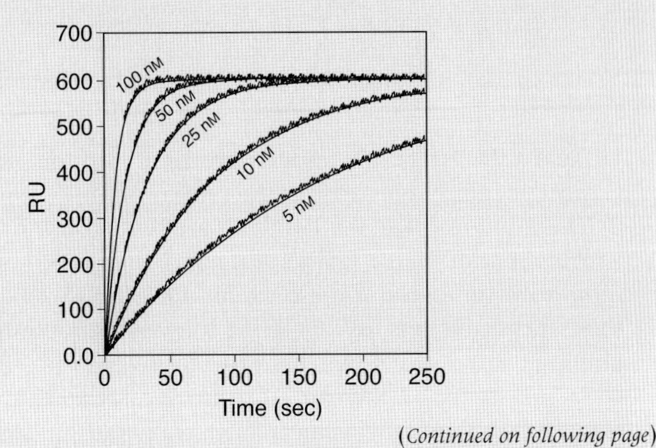

(*Continued on following page*)

Illustration 5.1. *(Continued)*

Next, you determine how the presence of Y affects the rate at which Q and X associate by measuring observed rate constants for the association of X with the QY complex (to ultimately give k'_1). To do so, you repeat the experiments described above, but with protein Y prebound to the immobilized DNA to form QY, and then flow protein X over the chip. Curves are obtained by fitting the data with Equation 17.

To determine k_1 and k'_1, you create new plots with the observed rate constants (k_{obs} or k'_{obs}) on the *Y* axis and the concentrations of protein X on the *X* axis and fit the two data sets using linear regression (see Equations 18 and 19). The slopes of the lines are equal to k_1 (1.4×10^6 M^{-1}s^{-1}) and k'_1 (7.2×10^6 M^{-1}s^{-1}), for the data monitoring QX (squares) and QXY (circles), respectively. The *Y* intercepts are too close to zero to allow k_{-1} and k'_{-1} to be determined from these lines. As can be seen in the plot and by comparing the resulting rate constants, you determine that k'_1 is approximately fivefold larger than k_1; hence, Y increases the rate at which X binds to the composite element on the DNA. You conclude that Y in part activates the XQ interaction by increasing the rate of association of X with Q. These results and your analysis provide insight into the mechanism for the cooperativity with which the QXY complex forms.

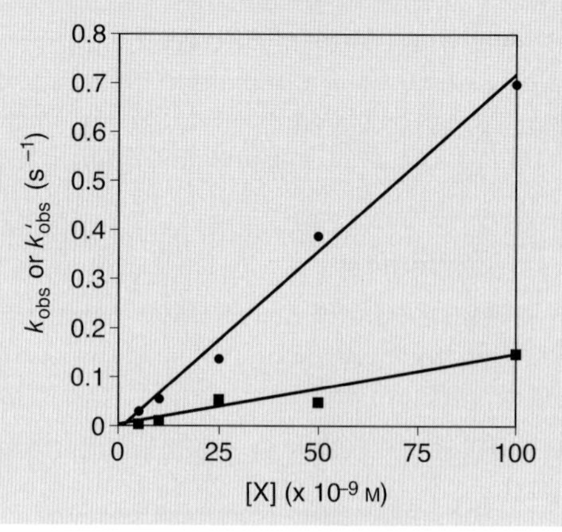

10^6 and 4.7×10^6 M^{-1}s^{-1}, respectively. Hence, in this example, molecule C causes a fourfold increase in the rate constant for association of A and B. The *Y* intercepts of the two lines are both equal to 0.01. It therefore appears that molecule C does not affect the rate constant for dissociation; however, because these values are close to zero, the accuracy of these measurements for k_{-1} and k'_{-1} is low. It is important to realize that although these measurements provide insight into the mechanism of cooperativity (i.e., C increases the rate of association between A and B), conclusions about the value of the cooperativity factor α cannot be made by comparing the values of k_1 and k_1' alone.

***Measuring* k_{-1} *and* k_{-1}'.** The information in the box below summarizes the key points for measuring k_{-1} and k_{-1}' for trimolecular systems, and the subsequent discussion expands on these points. Again, we recommend that you refer to Chapter 4 for a more detailed discussion of experimental design, including how to develop assays, block the association reaction, and design time courses.

The following equation describes the decay of AB complex over time

Measuring k_{-1} and k_{-1}'

1a. Develop an assay to monitor AB.
 b. Combine A and B and allow AB complex to form.
 c. Alter conditions to block new AB formation ($t = 0$).
 d. Measure the disappearance of AB over time.
 e. Plot data and determine k_{-1} (using Equation 20).
 f. If artifactual decay is suspected, k_{-1} is an upper limit.
2a. Develop an assay to monitor ABC.
 b. Combine AC and B and allow ABC complex to form.
 c. Alter conditions to block new ABC formation ($t = 0$).
 d. Measure the disappearance of ABC over time.
 e. Plot data and determine k_{-1}' (using Equation 21).
 f. If artifactual decay is suspected, k_{-1}' is an upper limit.
3. Compare k_{-1}' to k_{-1}.

and can be used to determine k_{-1}:

$$AB_t = (AB_{max} - AB_{min})(e^{-k_{-1}t}) + AB_{min} \qquad (20)$$

AB_t is the amount of AB complex at each time point, quantitated in a manner specific to the assay (e.g., phosphorimager units). The term AB_{max} defines the true starting point of the decay curve (the Y intercept of the curve at $t = 0$) that is obtained from fitting the data with Equation 20. This value of AB_{max} may differ from the AB_t experimentally measured at $t = 0$ (for further explanation, see Chapter 4). The term AB_{min} defines the plateau point of the decay curve. This equation takes into account (1) the possibility of a rapid decrease in the amount of AB complex on blocking new association (usually due to the presence of nonspecific complexes) and (2) the possibility that a portion of the AB complex dissociates very slowly or not at all (again, see Chapter 4 for clarification).

k_{-1}' is measured using a similar approach, with one key exception. In this case, molecule C is included in the reactions at a concentration that allows it to saturate all of the A in the reaction; therefore, the decay of the ABC complex is monitored. The following equation can be used to determine the k'_{-1}:

$$\text{ABC}_t = (ABC_{max} - ABC_{min})(e^{-k'_{-1}t}) + ABC_{min} \quad (21)$$

where ABC_t is the amount of the ABC complex at each time point. The terms in this equation are analogous to those in Equation 20. k'_{-1} is the rate constant for the dissociation of B from the ABC complex (see Equation 15). Depending on the assay used to monitor the decay of ABC, however, it may not be possible to distinguish between the dissociation of B from the AC complex (k'_{-1}) or the dissociation of C from the AB complex (k'_{-2}) (refer to the model in Equation 2 or 3). Whether B or C dissociates first from the ABC complex depends on the relative values of k'_{-1} and k'_{-2} for the reaction. If the experimental system does not allow the two pathways to be distinguished, then the rate constant measured using Equation 21 could be a composite of k'_{-1} and k'_{-2} and therefore greater than either k'_{-1} or k'_{-2}. In this situation, the measured rate constant for dissociation should be presented as an upper limit.

To determine k_{-1} and k'_{-1}, AB_t and ABC_t values are plotted, respectively, on the Y axis with time plotted on the X axis. A curve-fitting program (see Chapter 7) can be used to simultaneously solve for k_{-1}, AB_{max}, and AB_{min} in the case of Equation 20, and for k'_{-1}, ABC_{max}, and ABC_{min} in the case of Equation 21. In situations where it is likely that a rapid initial dissociation has occurred, it is best not to include the zero time point in the data set used for the curve fitting.

The plot in Figure 5.3 provides examples of how Equations 20 and 21 can be used to fit data. In these examples, the AB and ABC complexes were measured using phosphorimagery, and phosphorimager units were

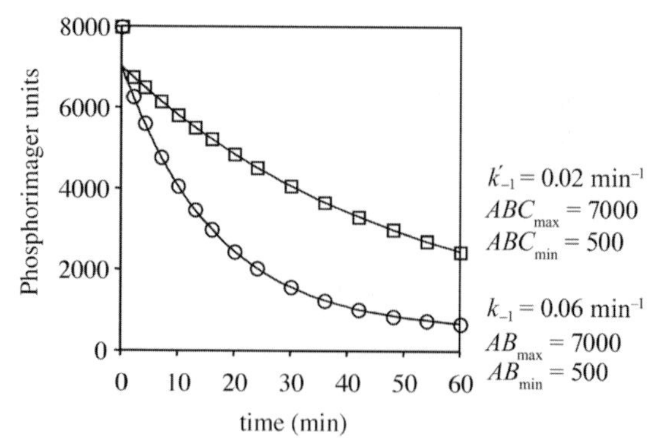

FIGURE 5.3. Decay curves for the dissociation of AB and ABC. The dissociation of AB (*circles*) and ABC (*squares*) were monitored using phosphorimagery (e.g., A was radioactively labeled) and the phosphorimager units were plotted versus time. The parameters used along with Equations 20 and 21 to generate the data points and the curves are shown.

plotted versus time. In the absence of molecule C, Equation 20 was used to generate the data points (circles) and curve using the following values: a k_{-1} of 0.06 min^{-1}, an AB_{max} of 7000 phosphorimager units, and an AB_{min} of 500 phosphorimager units. In the presence of molecule C, Equation 21 was utilized to generate the data points (squares) and curve using the following values: a k_{-1} of 0.02 min^{-1}, an AB_{max} of 7000 phosphorimager units, and an AB_{min} of 500 phosphorimager units. Thus, molecule C causes a threefold decrease in the rate constant for dissociation of B from A (from 0.06 to 0.02 min^{-1}), thereby stabilizing the complex. As was true for the association rate constants, knowing k_{-1} and k'_{-1} provides information on the mechanism of the cooperativity with which the ABC complex forms (i.e., C increases the kinetic stability of the AB complex), but it does not provide a measurement of the value of the cooperativity factor α.

Literature Example 5.1. Studying the mechanism by which Ca^{++} regulates the interaction between a protein and the cell membrane. *(Nalefski E.A. and Newton A.C. 2001. Biochemistry 40: 13216–13229.) Figures reprinted, with permission, from Nalefski and Newton (2001). (©American Chemical Society)*

In this Literature Example, the effect of a regulator on a binding reaction is studied in terms of cooperativity. The authors investigated the interaction between the C2 domain of protein kinase C and cellular membranes as regulated by Ca^{++}. Protein kinase C (PKC) is a cytosolic protein that binds Ca^{++} in its C2 domain and diacylglycerol in its C1 domain. Binding of these regulators drives PKC to interact with the cellular membrane, which activates its kinase activity and downstream signaling. Removal of the regulatory molecules (Ca^{++} and diacylglycerol) disrupts the interaction between PKC and the cell membrane. The authors of this paper sought to determine how Ca^{++} regulates membrane docking by PKC. Specifically, the experiments described below determined how Ca^{++} affected the dissociation rate constant for the membrane–C2 domain complex.

First, the authors developed an assay to determine the rate at which the C2 domain dissociated from phospholipid vesicles (which mimic the cell membrane) in the presence of Ca^{++} (i.e., the regulator). Purified C2 domain was preincubated with both Ca^{++} and vesicles to form a ternary complex (analogous to the ABC complex in Equation 2). The membrane vesicles contained small amounts of dansylated phospholipids. When the Ca^{++}-bound C2 domain formed a complex with the vesicles, fluorescence resonance energy transfer (FRET) occurred between tryptophans in the C2 domain and the dansylated phospholipids in the vesicles. To measure the rate at which Ca^{++}-bound C2 domain dissociated from the vesicles, at t = 0, ternary complexes were rapidly mixed with a solution containing a large molar excess of vesicles that did not contain dansylated phospho-

(Continued on following page)

Literature Example 5.1. *(Continued)*

lipids to trap the C2 domain after it dissociated. Loss in the FRET signal as the complexes decayed over time was monitored using a stopped-flow fluorescence spectrophotometer. The data were plotted as the relative dansyl emission versus time and fit with Equation 21:

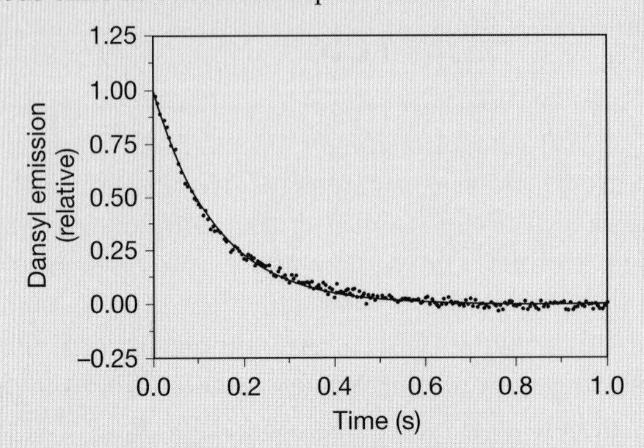

The rate constant for dissociation of the C2 domain from vesicles in the presence of Ca^{++} was 9.2 s^{-1}, which was independent of the concentration of trapping vesicles.

Determining the dissociation rate constant for the C2 domain–vesicle complex in the absence of Ca^{++} was not straightforward because the complex will not form in the absence of Ca^{++} (i.e., the C2 domain docking to membranes is dependent on Ca^{++}). Therefore, a "Ca^{++}-trapping" approach was used that effectively allowed dissociation of C2 domains from vesicles to be studied in the absence of Ca^{++}. Complexes containing Ca^{++}-bound C2 domain and vesicles were formed. At $t = 0$, complexes were mixed with a large molar excess of EDTA to rapidly trap free Ca^{++} present on mixing and after dissociation from the C2 domain. The dissociation of the Ca^{++}-free C2 domains from vesicles was then monitored over time. Plotting the loss in protein to membrane FRET over time and fitting the data with Equation 20 yielded a dissociation rate constant of 156 s^{-1}.

This Ca^{++}-trapping technique allowed measurement of the rate of membrane–C2 domain dissociation after Ca^{++} was sequestered. From these experiments, the authors were able to conclude that the presence of Ca^{++} stabilizes the ternary complex by causing a 17-fold decrease in the rate constant for dissociation between C2 domains and vesicles.

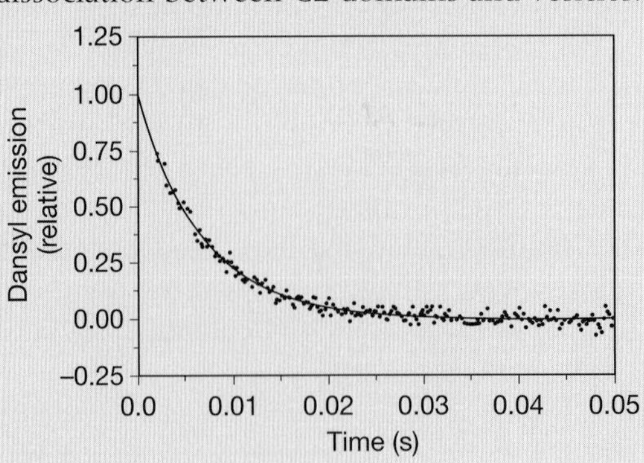

CHAPTER 6

Rate Constants for a Two-step Binding Reaction

IN THIS CHAPTER, WE DISCUSS APPROACHES to measuring rate constants for binding reactions that have two steps. We do not consider all possible two-step binding reactions, but limit our focus to a specific type of reaction found in multiple biological systems, namely, two-step bimolecular binding reactions in which the association of two molecules is followed by a transformation that does not involve an additional binding step. Understanding the rate constants that describe both the binding step and the transformation step(s) can reveal the rate-limiting step of the reaction. The intent of this chapter is to provide sufficient theory and equations such that rate constants governing this type of binding reaction can be measured. It will be clear from the ensuing discussion, however, that this two-step binding reaction is complicated, and obtaining rate constants for individual steps often requires that assumptions be made. Throughout the chapter, literature examples are provided to demonstrate practical aspects of measuring rate constants. Key equations are contained in boxes and are derived in Appendix 1. In addition, a computer simulation is provided on the Web site (http://kinetics.cshl.edu) that can be manipulated to help visualize concepts presented in the chapter.

Topics covered:

- Measuring rate constants for two-step bimolecular interactions
- Deciding whether to model a binding reaction as one step or two steps
- Regulation of a two-step binding interaction

TERMS AND PRINCIPLES

Rate constants for binding reactions are numeric representations of the time that it takes for molecules to associate or dissociate. In general, rate = (rate constant)(concentrations).

Two-step bimolecular binding reactions considered in this chapter occur when two molecules interact via a two-step mechanism consisting of an initial association followed by a second step that does not involve the binding of a third molecule. This two-step reaction can be expressed as A + B ↔ AB ↔ AB*, where AB and AB* are different from each other (e.g., in conformation), but identical in molecular composition. The formation of AB* involves (1) the initial binding of A to B, which is described by an association rate constant k_1 (units $\text{M}^{-1}t^{-1}$, where t is time) and a dissociation rate constant k_{-1} (units t^{-1}), and (2) a transformation after the initial binding that converts AB to AB*, which is governed by the forward and reverse rate constants k_2 (units t^{-1}) and k_{-2} (units t^{-1}), respectively. Most often, the unit of time used for expressing rate constants for biomolecular interactions is seconds; hence, we express rate constants in units of $\text{M}^{-1}\text{s}^{-1}$ or s^{-1} throughout this chapter.

Microscopic rate constants, as referred to in this chapter, are the actual rate constants for the individual steps in a two-step binding reaction. For example, the forward rate constants for the formation of AB (k_1) and the transformation to AB* (k_2) are the microscopic rate constants, whereas the experimentally measured value for formation of AB* from A + B is an observed rate constant.

Observed rate constants **(k_{obs})**, as discussed in this chapter, are experimentally measured values that describe the formation of AB* from A + B or the decay of AB* to A + B. Observed rate constants mathematically relate to the actual rate constants for association (k_1), dissociation (k_{-1}), or molecular transformations (k_2 and k_{-2}). k_{obs} values have units of s^{-1}.

TWO-STEP BIMOLECULAR INTERACTIONS (A + B ↔ AB ↔ AB*)

The formation of complexes containing two different molecules can involve one or more steps that occur before or after the binding step. The additional step(s) does not involve new binding, but instead consists of a transformation (e.g., a conformational change) in one of the two molecules or in the bimolecular complex. The derivations of equations that relate observed rate constants to microscopic rate constants for two-step reactions are complicated and require assumptions to be made about the relative values of the microscopic rate constants. Moreover, the equations for different types of two-step reactions differ from one another. Given this degree of complexity, we limit our focus in this chapter to a two-step reaction that involves the initial formation of a kinetically unstable complex, which then, in a second step, transforms into a kinetically stable complex. This type of reaction occurs frequently in biology, and its description provides an example of how to approach the study of other complex two-step binding reactions. Much of the discussion draws on information presented in Chapter 4 (rate constants for bimolecular complexes).

Theory and Equations

A two-step bimolecular binding reaction in which an initial binding event is followed by a transformation in the complex can be modeled by the following expression:

$$A + B \underset{k_{-1}}{\overset{k_1}{\rightleftharpoons}} AB \underset{k_{-2}}{\overset{k_2}{\rightleftharpoons}} AB^* \qquad (1)$$

where AB and AB* are complexes that have identical molecular compositions but are different from each other in structure or conformation. The formation of AB involves the binding of one molecule of A to one molecule of B, which is governed by an association rate constant k_1 with units of $\text{M}^{-1}\text{s}^{-1}$ and a dissociation rate constant k_{-1} with units of s^{-1}. The formation of AB* involves an intramolecular transformation of the AB complex. This transformation is governed by a forward rate constant k_2 and a reverse rate constant k_{-2}, both with units of s^{-1}. Note that we do not refer to k_2 and k_{-2} as association and dissociation rate constants, because the conversions between AB and AB* do not involve the binding of additional molecules.

Changes in the AB* complex over time can be described by the following rate equation (for a review of how to obtain this equation, see "General Rate Theory" in Chapter 4):

$$\frac{\text{change in [AB*]}}{\text{change in time}} = k_2[\text{AB}] - k_{-2}[\text{AB*}] \qquad (2)$$

Here, [AB] and [AB*] are molar concentrations. Note that this rate equation describes the accumulation of the final product over time, but it does not include [A], [B], k_1, or k_{-1}. The following rate equation that describes the change in [AB] over time contains [A] and [B], as well as all four rate constants:

$$\frac{\text{change in [AB]}}{\text{change in time}} = k_1[\text{A}][\text{B}] - k_{-1}[\text{AB}] - k_2[\text{AB}] + k_{-2}[\text{AB*}] \qquad (3)$$

The four rate constants cannot be measured directly. Rather, an observed rate constant (k_{obs}) can be measured, which describes the exponential accumulation of AB* after mixing A with B. The following equation relates the formation of AB* over time to an observed rate constant:

$$\frac{[AB^*]_t}{[AB^*]_{max}} = 1 - e^{-k_{obs}t} \tag{4}$$

Here, $[AB^*]_t$ is the concentration of AB* at each time point and $[AB^*]_{max}$ is the concentration of AB* when the reaction reaches equilibrium. k_{obs} is a measured rate constant with units of s^{-1} that will change as $[B]_i$ changes. The observed rate constant mathematically relates to the microscopic rate constants for the reaction. In the discussion below, we define and explain the mathematical expression for k_{obs} for a specific type of two-step reaction. The expression is different from that for one-step reactions presented in Chapter 4. In a later section, "Deciding Whether to Model a Binding Reaction as One Step or Two Steps," we relate these equations to one another.

Relating k_{obs} to microscopic rate constants for reactions where $k_{-1} >> k_2 >> k_{-2}$. To obtain an equation that relates k_{obs} to the microscopic rate constants in Equation 1, we make two assumptions and one experimental limitation. First, for many bimolecular biological complexes that form in two steps, the final functional complex is kinetically stable (i.e., the AB* complex forms more rapidly than it decays). In the two-step model shown in Equation 1, this would occur when $k_2 >> k_{-2}$; in other words, the formation of AB* from AB is more rapid than the decay of AB* back to AB. Second, in some two-step biological reactions, the AB complex is in rapid equilibrium with free A and B. This can occur, for example, when two molecules first bind relatively unstably to form an AB complex, which then transforms in some way to the kinetically stable and functional complex AB*. In the two-step model in Equation 1, this would occur when $k_{-1} >> k_2$; the dissociation of AB is more rapid than the transformation of AB to AB*. Together, the two assumptions made above lead to the condition $k_{-1} >> k_2 >> k_{-2}$. Third, reactions should be performed under conditions where the initial concentration of B ($[B]_i$) is much greater than the sum of the final concentrations of AB and AB* (for further explanation of this condition, see Chapter 4). The simplest way to achieve this condition is to set $[B]_i >> [A]_i$, ensuring that the total concentration of B at any point in a reaction is approximately equal to $[B]_i$.

Under the conditions described above, the relationship between k_{obs}, $[B]_i$, and the microscopic rate constants is (for derivation, see Appendix 1):

$$\boxed{k_{obs} = \frac{k_1[B]_i k_2}{k_1[B]_i + k_{-1}}} \tag{5}$$

This equation describes a hyperbola with k_{obs} values plotted on the *Y* axis and $[B]_i$ on the *X* axis. An example of this type of curve is shown in Figure 6.1. Although this equation relates k_{obs} to the microscopic rate constants, in this form, values for k_1, k_{-1}, and k_2 still cannot be obtained from experimental data.

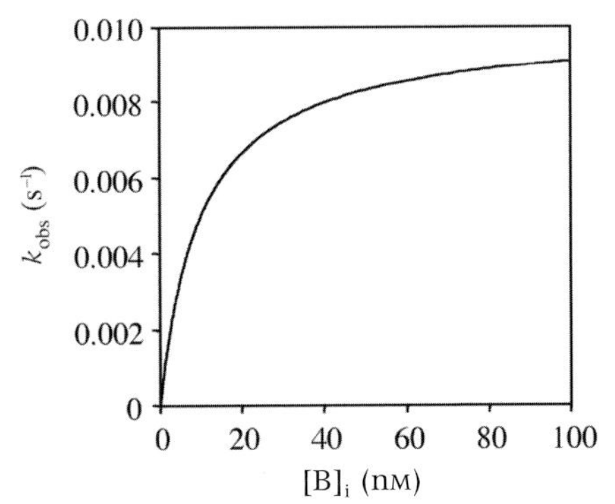

FIGURE 6.1. Curve illustrating how k_{obs} varies with the concentration of B_i for a two-step reaction. The curve was generated using Equation 5 and the following values: $k_1 = 1 \times 10^8$ M^{-1}s^{-1}, $k_{-1} = 1$ s^{-1}, and $k_2 = 0.01$ s^{-1}.

To make Equation 5 more useful, it can be transformed by taking the reciprocals of the terms on both sides of the equation (i.e., dividing the terms on each side of the equation into 1). After rearrangement, a new equation results (for derivation, see Appendix 1):

$$\frac{1}{k_{obs}} = \left(\frac{k_{-1}}{k_1 k_2}\right)\frac{1}{[B]_i} + \frac{1}{k_2} \qquad (6)$$

This equation describes a line. When $1/k_{obs}$ is plotted on the *Y* axis and $1/[B]_i$ on the *X* axis, the data points will fit a line with a slope equal to the term in parentheses and a *Y* intercept equal to $1/k_2$. This type of plot is referred to as a double reciprocal plot, an example of which is shown in Figure 6.2 (see also Simulation S6).

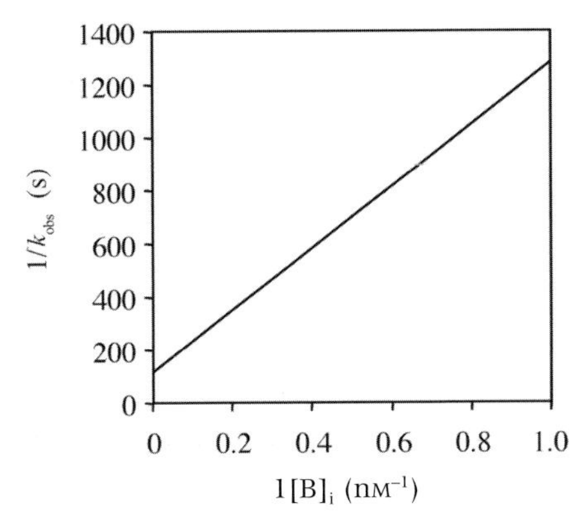

FIGURE 6.2. Double reciprocal plot of $1/k_{obs}$ versus $1/[B]_i$. The curve was generated using Equation 6 and the following values: $k_1 = 1 \times 10^8$ M^{-1}s^{-1}, $k_{-1} = 1$ s^{-1}, and $k_2 = 0.01$ s^{-1} (identical to those used to generate the curve in Fig. 6.1).

The term in parenthesis in Equation 6 has added meaning. Recall that the equilibrium dissociation constant for a binding reaction (K_D) is equal to the dissociation rate constant divided by the association rate constant (k_{-1}/k_1; for further explanation of K_D, see Chapter 2). Therefore, K_D can be substituted into Equation 6 to yield:

$$\frac{1}{k_{obs}} = \left(\frac{K_{D1}}{k_2}\right)\frac{1}{[B]_i} + \frac{1}{k_2} \tag{7}$$

where K_{D1} is the equilibrium dissociation constant (units of M) for the formation of AB (the first step of the two-step reaction). In Equation 7, k_2 is equal to the reciprocal of the *Y* intercept and K_{D1} is equal to the slope divided by the *Y* intercept. One way to think about the *Y* intercept of this double reciprocal plot is that as $[B]_i$ approaches infinitely high concentrations (i.e., $1/[B]_i$ approaches zero), the forward reaction cannot go any faster than the rate of the second transformation step, which is defined by k_2. Hence, the overall rate for the formation of AB* at high $[B]_i$ is limited by the second step, which does not directly depend on the concentration of B.

Dissociation rate constants for two-step reactions. Recall from the discussion in Chapter 4 that measuring dissociation rate constants for one-step bimolecular interactions involves three steps: (1) forming the complex, (2) changing the reaction conditions to prevent new association, and (3) measuring the decrease in complex over time. The second step is critical, because eliminating the association reaction allows dissociation to be studied in the absence of reassociation.

Measuring the rate of decay of AB* is accomplished via a similar approach: AB* is allowed to form, new association of A with B is blocked, and then decay of AB* is monitored over time. The rate constant measured is an observed rate constant with units of s^{-1}. To avoid confusion with the observed rate constant for the formation of AB* (k_{obs}), we refer to the observed rate constant for the decay of AB* as k_{off} in the remainder of this chapter. The following equation describes the decay of AB* over time:

$$\frac{[AB^*]_t}{[AB^*]_i} = e^{-k_{off}t} \tag{8}$$

where $[AB^*]_t/[AB^*]_i$ is the concentration of the AB* complex at each time point ($[AB^*]_t$) divided by the initial concentration of the AB complex ($[AB^*]_i$) in the reaction just before the point at which new association was blocked ($t = 0$). Equation 8 describes a curve that starts at 1 and exponentially decreases to 0 as time approaches infinity (see Chapter 4, Fig. 4.2). Importantly, under the limiting conditions described in the previous section ($k_{-1} >> k_2 >> k_{-2}$), $k_{off} = k_{-2}$, as derived in Appendix 1.

Experimental Considerations

In the following sections, we describe how to measure k_{obs} and k_{off} and how to determine K_{D1} and k_2 for two-step binding reactions where $k_{-1} >> k_2 >> k_{-2}$. We rely on many of the experimental details and equations that were covered in Chapter 4, so it may be helpful to refer back to this chapter.

Although this discussion focuses on two-step reactions, it is important to note that k_{obs} and k_{off} can be measured as described below for any bimolecular interaction, even if it is not known how many steps a binding reaction involves. Measuring k_{obs} and k_{off} can reveal whether the reaction occurs in two steps, and if it does, whether some of the microscopic rate constants can be determined. The process of distinguishing between one-step and two-step binding reactions using observed rate constants is discussed in a later section.

Summary: How to assess the kinetics of two-step reactions

Formation of AB*

1. Develop an assay to monitor AB*.
2. Combine A and B at $t = 0$ ($[B]_i >> [A]_i$).
3. Measure the accumulation of AB* over time.
4. Plot data and determine k_{obs} (using Equation 9).
5. Repeat Steps 2–4 at multiple different $[B]_i$.
6. Plot $1/k_{obs}$ versus $1/[B]_i$ and determine K_{D1} and k_2 (using Equation 7).

Decay of AB*

1. Develop an assay to monitor AB*.
2. Combine A and B and allow AB* complex to form.
3. Alter conditions to block new AB formation ($t = 0$).
4. Measure the disappearance of AB* over time.
5. Plot data and determine k_{off} (using Equation 10).
6. $k_{off} = k_{-2}$ (if $k_{-1} >> k_2 >> k_{-2}$).

Measuring* k_{obs} *for formation of AB*. Recall that the initial concentration of B in the reaction ($[B]_i$) must be much greater than the sum of the final concentrations of AB and AB*. The simplest way to achieve this experimentally is to set the concentration of B well above the concentration of A in the reactions. To measure k_{obs} for two-step binding reactions, an assay must be developed that accurately detects the formation of AB* over time. For many two-step reactions, the formation of AB* from AB will be relatively slow (i.e., k_2 is less than 0.1 s^{-1}), therefore manual sampling is often appropriate to monitor the formation of AB* over time. If the reaction occurs too fast to sample manually, an automated method will be needed to measure k_{obs}.

When the amount of AB* is measured, use the following equation to determine k_{obs}:

$$AB^*_t = AB^*_{max}\,(1 - e^{-k_{obs}t}) \tag{9}$$

where AB^*_t values are the amounts of AB* complex present at the different time points, quantitated in a manner specific to the assay (e.g., phosphorimager units). AB^*_{max} is the maximum amount of AB complex that forms as the reaction approaches equilibrium; it will have the same units as AB^*_t.

AB^*_t values are plotted on the *Y* axis and time on the *X* axis. k_{obs} and AB^*_{max} can be obtained by fitting the data with nonlinear regression using Equation 9 (see Chapter 7). For an example of a curve of this type, see Chapter 4, Figure 4.10. This process should then be repeated at multiple concentrations of $[B]_i$ to obtain multiple k_{obs} values.

Once k_{obs} is measured at multiple concentrations of B, a double reciprocal plot of $1/k_{obs}$ versus $1/[B]_i$ should be generated. As we described earlier, and as shown in Equation 7, when $k_{-1} >> k_2 >> k_{-2}$ the relationship between $1/k_{obs}$ and $1/[B]_i$ is described by a line with a slope equal to K_{D1}/k_2 and a *Y* intercept equal to $1/k_2$. Dividing the slope by the *Y* intercept yields K_{D1}. The values for the slope and *Y* intercept for a data set can be determined by using linear regression (see Chapter 7). An example of this type of plot is shown in Figure 6.2 (see also Simulation S6).

Measuring $\mathbf{k}_{off}$ ***for the decay of AB*.*** To measure k_{off}, an assay is needed to monitor the decay of the AB* complex over time. The general experimental outline is to (1) allow the maximum amount of AB* complex to form, (2) add a reagent to block new association of A with B, and (3) monitor the decay of the AB* complex over time. Approaches to blocking the association of A with B and defining a time course appropriate for the experiment are described in Chapter 4. The following equation should be used to determine the k_{off}:

$$AB^*_t = (AB^*_{max} - AB^*_{min})(e^{-k_{off}t}) + AB^*_{min} \tag{10}$$

Here, AB^*_t is the amount of AB* complex present at each time point, quantitated in a manner specific to the assay. The term AB^*_{max} defines the true starting point of the decay curve fit by the majority of the data (i.e., the *Y* intercept of the curve at $t = 0$ obtained from fitting the data to Equation 10). This value may differ from the value for AB^*_t experimentally measured at $t = 0$ (for further explanation, see Chapter 4). The term AB^*_{min} defines the point at which the decay curve fit by the majority of the data levels out.

When AB^*_t values are plotted on the Y axis and time on the X axis, values for k_{off}, AB^*_{max}, and AB^*_{min} can be obtained by fitting the data with a nonlinear regression program using Equation 10. In situations where it is likely that a rapid initial dissociation has occurred, it is best not to include the zero time point in the data set used for the computer fitting. (For an example of a plot fit with an equation that is similar to Equation 10, see Chapter 4, Fig. 4.8). For reactions where $k_{-1} >> k_2 >> k_{-2}$, $k_{off} = k_{-2}$.

Literature Example 6.1. Determining values for K_{D1} and k_2 at two promoters to understand differences in their frequencies of initiating transcription.

(McClure W.R. 1980. Proc. Natl. Acad. Sci. 77: 5634–5638.) Figure reprinted, with permission, from McClure (1980).

Initiation of transcription by RNA polymerase at bacterial promoters is an important control point for gene expression in *Escherichia coli.* Binding of RNA polymerase (R) to promoter DNA (P) occurs through a two-step mechanism of the type described in this chapter.

$$R + P \underset{k_{-1}}{\overset{k_1}{\rightleftharpoons}} RP_C \underset{k_{-2}}{\overset{k_2}{\rightleftharpoons}} RP_O$$

In the first step, RNA polymerase binds to promoter DNA to form what is referred to as a closed complex (RP_C) containing the enzyme and the DNA. In the second step, the closed complex undergoes an internal transformation to form an open complex (RP_O), during which the DNA unwinds to form the bubble required for transcription to initiate. This reaction occurs under the limiting condition $k_{-1} >> k_2 >> k_{-2}$; therefore, Equation 7 applies. In other words, closed complexes are in rapid equilibrium with free RNA polymerase and promoter DNA and then undergo a slower transformation into open complexes, which is rate-limiting for the reaction at high concentrations of polymerase. The study presented here used kinetic experiments to determine values for K_{D1} and k_2 at two different promoters of bacteriophage T7 (the D and A2 promoters) to reveal why these promoters initiate transcription with different frequencies.

The authors used an assay known as abortive initiation to monitor the formation of open complexes and to determine k_{obs} values at different polymerase concentrations. Abortive initiation is an enzymatic assay that takes advantage of the ability of RNA polymerase in open complexes (but not closed complexes) to abortively synthesize short RNA products 3 nucleotides in length when provided with ribonucleotides of the correct sequence. The 3-nucleotide RNAs are synthesized and released by the polymerase at a steady

(Continued on following page)

Literature Example 6.1. *(Continued)*

rate. Abortive initiation can be used to measure the average time required for RNA polymerase to form open complexes. When polymerase and promoter DNAs are mixed and abortive initiation is monitored, a lag occurs en route to reaching the linear steady-state synthesis and release of 3-nucleotide RNA. This lag time (referred to as τ_{obs}) can be determined experimentally and is equal to $1/k_{obs}$. The assay used to measure k_{obs} in this study is quite different from directly monitoring the accumulation of AB* over time, as we have described in this chapter. Nonetheless, the measured k_{obs} values can be used to determine K_{D1} and k_2 for the formation of open complexes at any promoter.

Using the abortive initiation assay, k_{obs} values were measured at multiple different RNA polymerase concentrations (always greater than the concentration of DNA) at two different promoters. Once the k_{obs} values were measured, values for both k_2 and K_{D1} were determined by generating double reciprocal plots. The data were plotted with $1/k_{obs}$ on the Y axis and 1/[RNA polymerase] on the X axis and fit with a line. The squares correspond to data obtained using the D promoter and the circles correspond to data obtained using the A2 promoter.

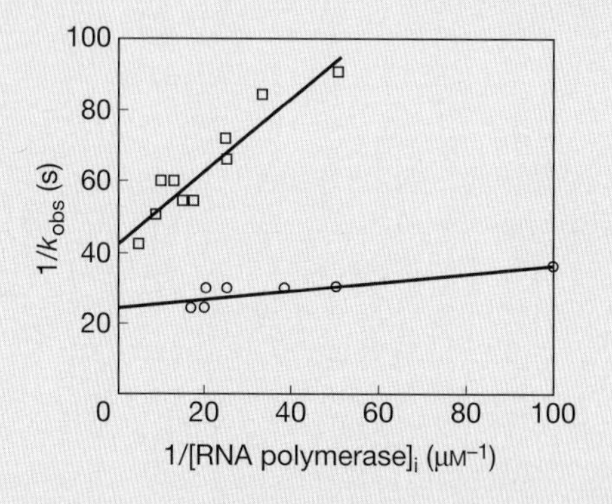

According to Equation 7, the Y intercepts are equal to $1/k_2$ and the slopes are equal to K_{D1}/k_2, allowing K_{D1} to be calculated. These values are shown in the table below.

A2 promoter		D promoter	
K_{D1} (nM)	k_2 (s^{-1})	K_{D1} (nM)	k_2 (s^{-1})
4.8	0.04	25	0.024

From this work, it is clear that the A2 promoter has both a lower K_{D1} and a greater k_2 than the D promoter. Therefore, both steps of the binding reaction between RNA polymerase and promoter DNA are more favorable at the A2 promoter than at the D promoter, explaining why the A2 promoter exhibits a greater frequency of initiation.

In some experimental systems, AB* and AB can be monitored simultaneously. The ratio of [AB*] to [AB] at equilibrium is equal to k_2/k_{-2}: The larger the ratio, the greater the difference between k_2 and k_{-2}. Therefore, if both complexes can be detected, the ratio of [AB*] to [AB] can be used to confirm the assumption that $k_2 >> k_{-2}$.

DECIDING WHETHER TO MODEL A BINDING REACTION AS ONE STEP OR TWO STEPS

It is often not known whether a binding reaction occurs with a single step, as described in Chapter 4, or with two steps, as explained in this chapter. Here, we discuss how to assess whether a one-step or a two-step model is most consistent with the k_{obs} measurements. The following discussion again relies on the assumption that, for two-step reactions, AB* is kinetically stable; therefore $k_2 >> k_{-2}$. It should be emphasized that the following analysis is useful for any binding reaction; the outcome of the analysis can sometimes indicate that the working model for a reaction should be reconsidered.

Recall that for one-step bimolecular interactions, a linear relationship exists when k_{obs} values are plotted versus $[B]_i$, as described in Chapter 4. This is because for one-step reactions:

$$k_{obs} = k_1[B]_i + k_{-1} \tag{11}$$

Hence, if k_{obs} values for a one-step reaction are measured at a variety of B concentrations and the data are plotted (with k_{obs} on the Y axis and $[B]_i$ on the X axis), the data should be linear. We refer to plots of k_{obs} versus $[B]_i$ as direct plots.

Double reciprocal plots can also be generated for one-step reactions. A double reciprocal transformation of Equation 11 results in an equation for a hyperbola that begins at zero and approaches $1/k_{-1}$ at very low concentrations of B (i.e., very high values of $1/[B]_i$):

$$\frac{1}{k_{obs}} = \frac{1/[B]_i}{k_1 + k_{-1}\,(1/[B]_i)} \tag{12}$$

Hence, if $1/k_{obs}$ values are plotted versus $1/[B]_i$ for a one-step reaction, the data should look curved when plotted.

Importantly, the shapes of the plots described by Equations 11 and 12 are the opposite of what is observed for two-step binding reactions. For two-step reactions, a direct plot of k_{obs} versus $[B]_i$ is curved (Equation 5), and a double reciprocal plot of $1/k_{obs}$ versus $1/[B]_i$ is linear (Equation 7). The four plots in Figure 6.3 illustrate the shapes of the ideal direct and double reciprocal plots for one-step and two-step reactions.

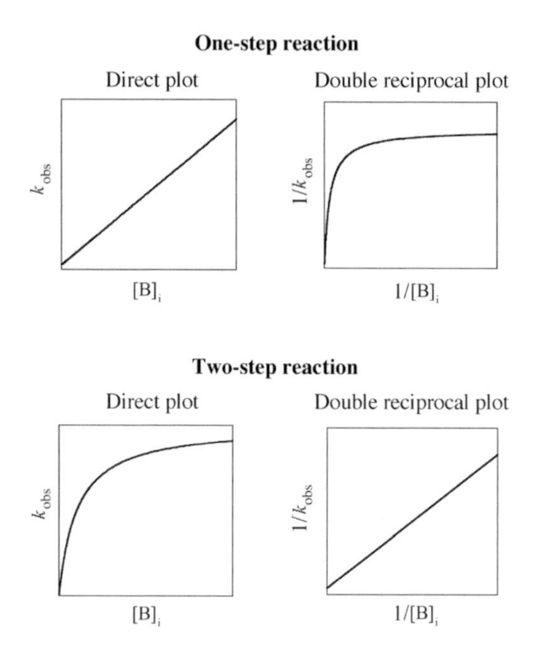

FIGURE 6.3. Plots demonstrating the shapes of the lines obtained for direct and double reciprocal plots of one-step and two-step binding reactions.

To assess whether a binding reaction is best modeled as one step or two steps, start by generating two plots of the experimental data: (1) a direct plot (k_{obs} vs. $[B]_i$) and (2) a double reciprocal plot ($1/k_{obs}$ vs. $1/[B]_i$). Fit the data points on both plots using the general equation for a line ($Y = mX + b$) to determine the slopes and Y intercepts of the lines that best fit the data. (Obviously, the fit will not be very good if the data are not linear.) For each of the plots, use the answers to the following two questions to assess whether the reaction should be modeled as one step or two steps: (1) Do the data fit a line (linear) or do the data curve toward a plateau in the upper right corner of the plot (curved)? (2) If the data appear linear, is the Y intercept of the line significantly greater than zero (>0) or is it equal to or close to zero (=0)?

Table 6.1 considers possible shapes of the double reciprocal and direct plots, as well as values for the Y intercept. An "na" is present in a Y intercept column whenever the plot is curved and a Y intercept does not have meaning. The two ideal and unambiguous situations are described in the first two rows. As shown in Row 1, when the double reciprocal plot is curved and the direct plot is linear with a Y intercept greater than zero, the reaction is best modeled as one step. As shown in Row 2, when the double reciprocal plot is linear with a Y intercept greater than zero and the direct plot is curved, the reaction is best modeled as two steps.

TABLE 6.1. Assessing Whether Binding Reactions are Best Modeled as One Step or Two Steps

	Double reciprocal plot ($1/k_{obs}$ vs. $1/[B]_i$)		Direct plot (k_{obs} vs. $[B]_i$)		
Row	Plot shape	Y intercept	Plot shape	Y intercept	Number of steps
1	Curved	na	Linear	> 0	1
2	Linear	> 0	Curved	na	2
3	Curved	na	Linear	= 0	1
4	Linear	= 0	Linear	≥ 0	1
5	Linear	> 0	Linear	= 0	2
6	Linear	= 0	Curved	na	2
7	Curved	na	Curved	na	2
8	Linear	> 0	Linear	> 0	1 or 2

The remaining rows in Table 6.1 show examples of curves that do not fit the ideal situations. Rows 3 and 4 provide examples of one-step reactions. In Row 3, observing a curved double reciprocal plot and a linear direct plot with a *Y* intercept of zero is most easily explained by a one-step reaction where the k_{-1} is very small and the *Y* intercept of the direct plot is indistinguishable from zero. Row 4 depicts a one-step reaction in which the double reciprocal plot appears linear (instead of curved) with a *Y* intercept of zero. This can occur if the concentrations of B used were all clustered at the beginning of a hyperbola; therefore, the data points did not plateau and the curve can be well simulated by a line. Thus, in situations where both plots appear linear, the double reciprocal plot has a *Y* intercept of zero, and the direct plot has a *Y* intercept greater than zero, the reaction is best modeled as one step. Note that, in this case, a *Y* intercept of zero for the double reciprocal plot does not prove that a reaction has only one step, but would be consistent with this model.

Rows 5 through 7 provide examples of two-step reactions. In Row 5, the direct plot appears linear (instead of curved) with a *Y* intercept of zero. Again, this can occur when the concentrations of B used were all clustered at the beginning of a hyperbola, where the curve can be well simulated by a line. Thus, if both plots appear linear, with the double reciprocal plot having a *Y* intercept greater than zero and the direct plot having a *Y* intercept equal to zero, the reaction is best modeled as two steps. In general, when $1/k_{obs}$ is plotted versus $1/[B]_i$ for a reaction, if the data approach a value greater than zero at high $[B]_i$, this would indicate that the binding reaction has more than one step. As indicated in Rows 6 and 7, whenever the direct plot appears curved, the reaction should be modeled as two steps. This is because the direct plot of a one-step reaction must be linear; there is no way that the direct plot of the data for a one-step reaction will reliably appear curved. Finally, if both lines are linear with intercepts greater than zero (as shown in Row 8), both one-step and two-step models should be considered.

Using the analysis described above, it should be possible to decide whether a binding reaction is best modeled as one step or two steps. For reactions that are best modeled as one step, microscopic rate constants can be determined as described in Chapter 4. For reactions that are best modeled as two steps, it is possible to obtain K_{D1} and k_2 values if the reaction fits the assumptions described in the earlier sections of this chapter.

REGULATORS OF TWO-STEP BINDING REACTIONS (ZA + B ↔ ZAB ↔ ZAB*)

Many biological binding reactions in nature are controlled by regulatory molecules. In addition, non-natural regulators of binding reactions are often identified or designed. We have discussed how regulators affect one-step binding reactions in previous chapters: Chapter 2 discusses competitive inhibitors and their effects on affinity constants for bimolecular interactions; Chapter 3 discusses how binding affinity can be affected by molecules acting with cooperativity, which can be thought of as noncompetitive regulation; and Chapter 5 discusses the effect of regulatory molecules on the association and dissociation rate constants for binding reactions (in the context of cooperativity). Here, we discuss an additional class of regulatory molecules that control the forward and reverse rates of multistep binding reactions, and in particular the second (transformation) step of two-step binding reactions. We limit the discussion to regulators of the type of reaction described earlier in the chapter, where $k_{-1} >> k_2 >> k_{-2}$. Therefore, the theory and equations we have already described apply in this section as well.

Theory and Equations

Regulators of two-step binding reactions. A regulator that affects the kinetics of a two-step binding reaction can be modeled as:

$$\begin{array}{ccccc} A + B & \underset{k_{-1}}{\overset{k_1}{\rightleftharpoons}} & AB & \underset{k_{-2}}{\overset{k_2}{\rightleftharpoons}} & AB^* \\ + & & + & & \\ Z & & Z & & \\ \Updownarrow & & \Updownarrow & & \\ ZA + B & \underset{{}_Zk_{-1}}{\overset{{}_Zk_1}{\rightleftharpoons}} & ZAB & \underset{{}_Zk_{-2}}{\overset{{}_Zk_2}{\rightleftharpoons}} & ZAB^* \end{array} \qquad (13)$$

where A and B are the two molecules that form a bimolecular complex via a two-step reaction, and Z is the regulatory molecule that can associate either with A or with the AB complex to affect the kinetics of one or both steps of the two-step binding reaction. (Note that unlike the description of competitive inhibitors in Chapter 2, Z does not compete with B for binding to A. Moreover, Z could also associate with the AB* complex, although this has been omitted from the model.) Rate constants are shown for the two-step reaction in the absence of Z (at the top) and the two-step reaction in the presence of Z (at the bottom, indicated with a "Z" subscript).

The vertical arrows indicate the association and dissociation of Z with A and AB. There are rate constants that govern these binding events; however, they are not important for the discussion that follows. This is because it is simplest to study the effects of a regulator on the kinetics of the AB-binding reaction under conditions where the regulator saturates A, which allows the ZA complex to be considered as a single entity. In this situation, Equation 13 can be considered as two separate two-step binding reactions: one in the absence of the regulator and one in the presence of the regulator:

$$\begin{gathered} A + B \underset{k_{-1}}{\overset{k_1}{\rightleftharpoons}} AB \underset{k_{-2}}{\overset{k_2}{\rightleftharpoons}} AB^* \\ ZA + B \underset{{}_Zk_{-1}}{\overset{{}_Zk_1}{\rightleftharpoons}} ZAB \underset{{}_Zk_{-2}}{\overset{{}_Zk_2}{\rightleftharpoons}} ZAB^* \end{gathered} \tag{14}$$

Assessing the effect of a regulator (Z) on the rate of formation of the AB* complex. Measuring the effect of Z on the kinetics of formation of the AB* complex requires measuring k_{obs} values in the absence and presence of Z. The theory and equations presented earlier for measuring rate constants for formation of AB* are also used to assess the effect of the regulator, and the same assumptions apply. Specifically, we consider two-step reactions in which the ZAB complex is in rapid equilibrium with free ZA and free B, and the dissociation of ZAB is more rapid than the transformation of ZAB to ZAB* (${}_Zk_{-1} >> {}_Zk_2 >> {}_Zk_{-2}$). In this situation, the equation relating ${}_Zk_{obs}$ to $[B]_i$ is:

$$\frac{1}{{}_Zk_{obs}} = \left(\frac{{}_Zk_{-1}}{({}_Zk_1)({}_Zk_2)}\right)\frac{1}{[B]_i} + \frac{1}{{}_Zk_2} \tag{15}$$

This is an equation for a line, with a slope equal to the term in parentheses and a Y intercept equal to $1/{}_Zk_2$; it is analogous to Equation 6. This equation can be converted to:

$$\frac{1}{{}_Zk_{obs}} = \left(\frac{{}_ZK_{D1}}{{}_Zk_2}\right)\frac{1}{[B]_i} + \frac{1}{{}_Zk_2} \tag{16}$$

where ${}_ZK_{D1}$ is the equilibrium dissociation constant for the ZAB complex. This equation is analogous to Equation 7.

Experimental Considerations

The approach to measuring k_{obs} in the presence of a regulator Z is identical to the description of how to assess the kinetics of two-step reactions presented earlier in this chapter. The only difference is that ZA complexes are formed first and are used in place of A. This discussion presumes that all of the A in the reaction is bound to Z. If this is not the case, then the measured effect of Z will be a partial effect and should therefore be considered as a lower limit. The brief discussion that follows will highlight possible effects that might be observed when studying the regulation of two-step reactions where $k_{-1} >> k_2 >> k_{-2}$ and ${}_Zk_{-1} >> {}_Zk_2 >> {}_Zk_{-2}$.

Measuring the effect of Z on the formation of AB*. When the amount of ZAB* is measured, the following equation should be used to determine k_{obs}:

$$\text{ZAB}^* = ZAB^*_{max}(1 - e^{-{}_Zk_{obs}t}) \quad (17)$$

ZAB^*_t values are plotted on the Y axis and time on the X axis. ${}_Zk_{obs}$ and ZAB^*_{max} can be obtained by fitting the data with nonlinear regression using Equation 17 (see Chapter 7). This process should then be repeated at multiple concentrations of $[B]_i$ to obtain different ${}_Zk_{obs}$ values. k_{obs} values should also be measured for the formation of AB* in the absence of Z.

Once k_{obs} and ${}_Zk_{obs}$ values are measured at multiple concentrations of B, double reciprocal plots of $1/k_{obs}$ versus $1/[B]_i$ and $1/{}_Zk_{obs}$ versus $1/[B]_i$ should be generated. Equations 7 and 16, which both describe lines, are used to determine the slopes and the Y intercepts in the absence and presence of Z, respectively.

Figure 6.4 shows how two different activators could affect the parameters governing the two-step binding reaction. The solid line represents the reaction in the absence of activators. One activator decreases K_{D1} (upper dashed line; ${}_ZK_{D1} < K_{D1}$), reflected by a change in slope with respect to the solid line. The second activator increases k_2 (lower dashed line; ${}_Zk_2 > k_2$), reflected by a change in the Y intercept with respect to the solid line.

Assessing whether a regulator is an activator or a repressor. Regulators of two-step binding reactions can affect the rate constants for the binding step, the intramolecular transformation step, or both steps. For a regulator that affects only one of the two steps, it is straightforward to decide whether it is an activator or a repressor of the binding reaction between A and B. If Z decreases K_{D1} (${}_ZK_{D1} < K_{D1}$) or increases k_2 (${}_Zk_2 > k_2$), then Z is an activator because the formation of the AB* complex becomes more favorable. If Z increases K_{D1} (${}_ZK_{D1} > K_{D1}$) or decreases k_2 (${}_Zk_2 < k_2$), then Z is an repressor because the formation of AB* becomes less favorable. In some cases, however, a regulator might activate one of the steps while repressing the other. Deciding whether such a regulator is an overall activator or an overall repressor of the two-step binding reaction requires that a sim-

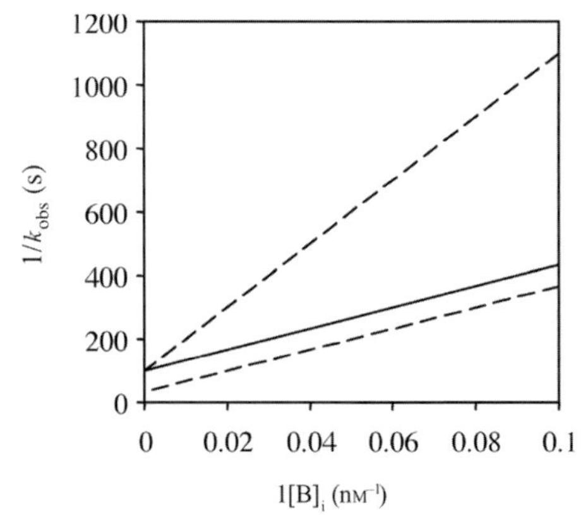

FIGURE 6.4. Double reciprocal plots of $1/k_{obs}$ versus $1/[B]_i$ in the absence and presence of activators. The solid line was generated using Equation 6 and the following values: $k_1 = 1 \times 10^7$ M^{-1}s^{-1}, $k_{-1} = 0.1$ s^{-1}, and $k_2 = 0.01$ s^{-1}. The upper dashed line depicts the effect of an activator that increases k_1 to 3×10^7 M^{-1}s^{-1}. The lower dashed line depicts the effect of an activator that increases k_2 to 0.03 s^{-1}.

ple calculation involving the equilibrium constant for the binding step and the rate constant for the intramolecular transformation step be performed. Specifically, the effect of the regulator on the quotient of k_2/K_{D1} should be determined. If the regulator increases this number (${}_Zk_2/{}_ZK_{D1} > k_2/K_{D1}$), then the regulator is an overall activator of the two-step binding reaction. If the regulator decreases this number (${}_Zk_2/{}_ZK_{D1} < k_2/K_{D1}$), then the regulator is an overall repressor of the two-step binding reaction.

Literature Example 6.2. Regulating the rate at which a translation initiation factor associates with the 5′ caps on mRNAs. *(Sha M., Wang Y., Xiang T., van Heerden A., Browning K.S., and Goss D.J. 1995. J. Biol. Chem. 270: 29904– 29909). Figures reprinted, with permission, from Sha et al. (1995). (© American Society for Biochemistry & Molecular Biology)*

Initiation of translation requires the assembly of a large protein–nucleic acid complex, ultimately leading to the recruitment of mRNA to the ribosome. In eukaryotes, mRNAs have a 5′ cap (m^{7}GpppX), which facilitates efficient translation as well as message stability. The 5′ cap is recognized by eukaryotic translation initiation factor 4F (eIF4F), and an isozyme form of eIF4F termed eIF-(iso)4F that exists in higher plants. eIF-(iso)4F binds to the cap via the two-step mechanism discussed in this chapter: an initial binding step in rapid equilibrium followed by a slow conformational change in the complex. eIF-(iso)4F is composed of two protein subunits, p28 and p86. These authors found that p28 contains the cap-binding activity and p86 decreases the rate at which p28 binds the cap. Therefore, in the following description, p86 is considered a regulator of the p28-cap complex.

(Continued on following page)

Literature Example 6.2. *(Continued)*

The forward reaction for forming the complex between p28 or eIF-(iso)4F (p28/p86) and the cap was monitored using stopped-flow fluorescence. p28 or eIF-(iso)4F was rapidly mixed with an excess of the m^7GpppG cap and binding was monitored over time by the increase in the fluorescence intensity of the cap, which was induced by association of either p28 or eIF-(iso)4F. The data were plotted with change in fluorescence intensity on the *Y* axis and time on the *X* axis, and fit to obtain k_{obs}. This was repeated at different concentrations of cap, all of which were greater than the concentration of protein complex. The following plot shows a representative k_{obs} measurement for the association of cap with eIF-(iso)4F.

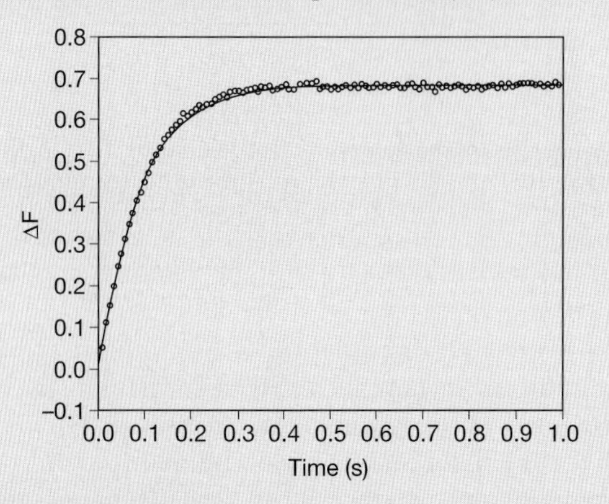

Next, double reciprocal plots of $1/k_{obs}$ versus 1/[cap] were generated and fit with a line (p28, triangles; eIF-(iso)4F, circles).

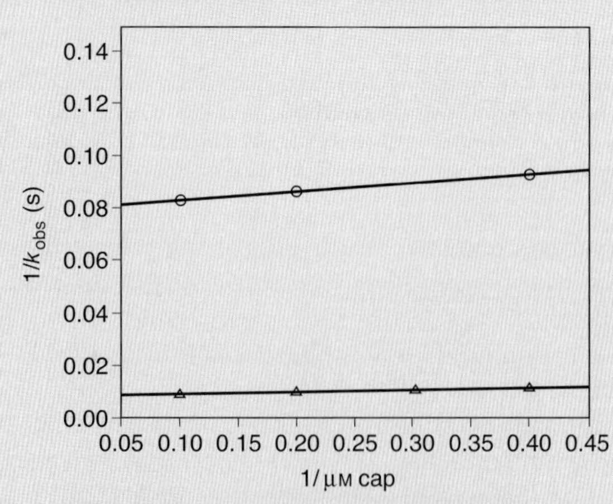

Values for k_2 were determined from the *Y* intercepts of the plots according to Equation 16. The k_2 values were 123.3 ± 8.6 s^{-1} for p28 and 12.2 ± 0.3 s^{-1} for eIF-(iso)4F, indicating that the isomerization (the second step of the binding reaction) occurs more slowly in the presence of the p86 subunit. Hence, p86 acts as a repressor.

CHAPTER 7

Data Analysis

THROUGHOUT THIS VOLUME THUS FAR, we have described how to obtain affinity constants and rate constants by plotting experimental data and fitting them with a specific equation. Here, we explain exactly how to fit data to curves defined by specific equations; this typically requires computer software. We cover the four types of curves discussed in this book: lines, binding affinity curves, association rate curves, and dissociation rate curves. In addition, we provide tables of sample data that can be used to practice and troubleshoot curve fitting using computer software. Finally, we include a discussion of assessing error in experimental measurements.

Topics covered:

- Fitting data to a line
- Fitting data to a hyperbolic binding affinity curve
- Fitting data to an exponential association rate curve
- Fitting data to an exponential dissociation rate curve
- Assessing error

TERMS AND PRINCIPLES

Regression is a technique for fitting an equation to a set of data points. Regression is used to determine the values for unknown variables in an equation such that the line or curve defined by the equation best fits through the experimental data points on a scatter plot. Linear regression draws a line to describe a linear relationship between variables. Nonlinear regression draws a curve to describe a nonlinear (e.g., exponential) relationship between variables. Many times, the phrase "curve fitting" is used in place of linear or nonlinear regression. A computer program is used to perform the regression analysis.

Initial values are provided by the investigator and used by computer programs to perform nonlinear regression. They are estimated values of the unknown variables that the computer will solve for while determining the curve that best fits the data set. Descriptions of how to estimate initial values are provided in the chapter.

95% confidence interval defines the range in which there is a 95% chance that the true value for a measured parameter falls. Consider a data set that when fit with the equation for a line has a calculated slope of 1 with a 95% confidence interval of 0.87 to 1.13. From this result, it can be stated with 95% confidence that these data form a line with a slope between 0.87 and 1.13. Alternatively, it could be stated that the slope is 1.0 ± 0.13, where the error represents the 95% confidence interval. Note that many programs will also calculate and display standard errors or other statistical parameters. Unless it is well understood how these parameters were calculated, they are not generally useful and should not be confused with 95% confidence intervals.

$\mathbf{R^2}$ provides a measure of how well the data fit a line or a curve. R^2 values for a fit can range from 1 to zero, with 1 indicating that the data fit the line or curve perfectly. As R^2 decreases, the line or curve fits the data less and less well. At some low R^2 value, which is often the judgment of the investigator, it must be concluded that the data are not fit by the equation for the line or curve.

CURVE FITTING

We explain here how to fit data to the four general types of curves used throughout this book: lines, hyperbolic binding affinity curves, exponential association rate curves, and exponential dissociation rate curves. Each section contains advice on how to plot data, fit data with the appropriate equation, and consider error in the results of the fit. Each section can be read independently of the others, with the following exception. The discussion of 95% confidence intervals and R^2 values is only included in the "Lines" section; however, the information here is applicable to all four types of curves. Importantly, three common themes emerge when considering error in the curve-fit: (1) More points on the curve result in a more accurate fit, (2) broadly distributed points result in a more accurate fit, and (3) the data points with the highest Y values dominate the fit. These themes are explained and an illustration is provided in the following sections.

In the examples that follow, we used the software Prism to perform regression, although many other regression programs also work well. For Prism, the box on the facing page provides a brief step-by-step guide to entering data, equations, and initial values for regression.

Using Prism Software

1. After starting the software, enter the XY graphing mode.
2. Enter data in the X and Y columns (note that multiple Y sets can be entered for a single X set).
3. Enter the "Analyze" tool.
4. Select "Liner regression" if fitting a line and then click the "OK" button. Click the "OK" button in the next window that appears. Go to Step 14.
5. Select "Nonlinear regression" if fitting curved data and then click the "OK" button.
6. Click on the "More equations" radio button.
7. If a previously entered equation is used, choose it from the list. If a new equation is entered, select "[Enter your own equation]." If the latter applies, then follow Steps 8–12 below.
8. Type a name for the equation.
9. Enter the equation (see the sections below for advice on what to enter).
10. Click on the "Rules for initial values" button.
11. Enter initial values for the variables (see sections below for advice).
12. Click the "OK" button.
13. Click the "OK" button on the next window that appears.
14. The values of the variables that Prism determines can be seen in the "Table of results."
15. A plot of the data points with the curve is available in the "Graphs" folder.
16. If necessary, the analysis parameters can be changed under the "Change" tool.

Lines

Fitting data. Fitting data with the equation for a line is straightforward and can be performed by many calculators and spreadsheet software programs. If a plotting program is used (such as Prism discussed above), the program will likely have a linear regression option that will fit linear data sets, thereby alleviating the need to enter the equation for a line. The general equation for a line is:

$$Y = mX + b \qquad (1)$$

Here, m is the slope and b is the Y intercept. As stated in Chapter 1, defining a line requires a minimum of four points and preferably five or more.

The higher* Y *values will dominate the fit. Experimentally measured data points will each have inherent error. It is important to understand how

TABLE 7.1. Linear Sample Data

X	Y	Y1	Y2	Y3	Y4	Y5
1	53	**42**	53	53	53	53
3	59	59	**47**	59	59	59
10	80	80	80	**64**	80	80
30	140	140	140	140	**112**	140
100	350	350	350	350	350	**280**
Slope	3.0	3.0	3.0	3.0	3.0	2.3
95% CI		2.9–3.2	2.8–3.3	2.7–3.4	2.4–3.6	1.9–2.7
***Y* intercept**	50	47	46	46	45	57
95% CI		38–56	36–57	31–60	19–71	39–75
R^2	1.000	0.999	0.999	0.997	0.990	0.992

error in the data can affect the values of the slope and *Y* intercept obtained from a linear curve fit. If each data point has a similar percentage error associated with it, then the error in the larger value data points (i.e., upper right-hand region of a plot) will more dramatically alter the outcome of the fit. This is true if the slope of the line is positive (as is the case for the types of lines described in this book).

This point is demonstrated in Table 7.1. The columns labeled "X" and "Y" contain the *X* and *Y* values for five data points that fit a perfect line. In each of the remaining columns (Y1–Y5), the *Y* value for a single data point was decreased by 20% (simulating 20% error), as shown in bold. The slopes and *Y* intercepts that result from fitting each of the data sets using linear regression are shown in the Y columns below the data. The 95% confidence intervals (95% CI) and R^2 obtained with each fit are also included in the table and are discussed below. As can be seen, only the error in the last point (i.e., the point having the largest *Y* value, which is shown in column Y5) causes a change in the slope compared to the perfect line (column Y). Moreover, error in this point also causes the greatest change in the *Y* intercept.

The perfect data (column Y) and the data containing error in the last point (column Y5) were plotted versus the *X* values (column X) and the data were fit using linear regression, as shown in Figure 7.1. The dashed line is the fit of the perfect data and the solid line is the fit of the data containing error in the last point. The difference in the slopes of the two lines can be clearly seen. It is important to note that the solid line appears to fit all five data points well. If this were experimental data, it would not likely be clear to the investigator that the last data point contained significant error. The best means of ensuring that accurate slopes and *Y* intercepts are obtained is to include as many data points as possible and

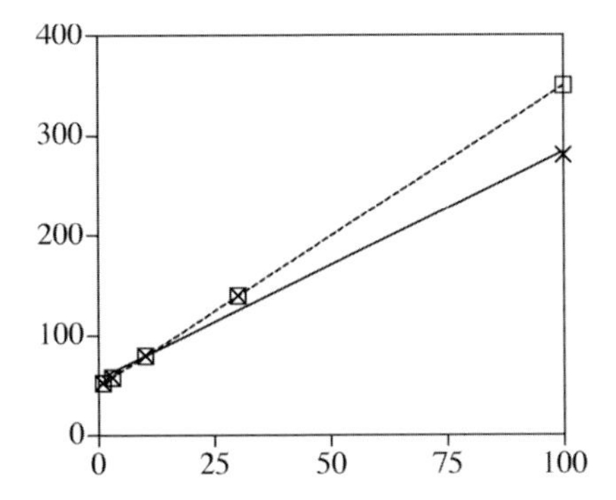

FIGURE 7.1. Illustration of how error in the point representing the highest *Y* value can influence the variables obtained from linear regression. The data plotted are from Table 7.1 with *X* values from column X and *Y* values from columns Y (*open squares*) and Y5 (*crosses*). The lines were obtained using linear regression, with the *dashed line* corresponding to the Y column data points and the *solid line* corresponding to the Y5 column data points. The slopes and *Y* intercepts of the two lines are shown in Table 7.1. Because this is a hypothetical illustration, the *X* and *Y* axes do not have units.

to make sure that they are distributed evenly along the plot (i.e., not all grouped in one area). Because the point with the highest *Y* value dominates the calculation of the slope and *Y* intercept, adding one or more additional data points to this region of the plot would help in obtaining more accurate measurements of the slope and *Y* intercept.

95% confidence intervals and $\mathbf{R^2}$**.** When fitting data to the equation for a line, many programs calculate the 95% confidence intervals associated with the slope and *Y* intercept. These intervals can be used as a measure of the error associated with the calculated slope and *Y* intercept. Table 7.1 lists the confidence intervals determined from curve fitting the plots of the data in columns Y1–Y5 versus the X column. Note that there are no 95% confidence intervals listed in the Y column, because these are perfect data. The 95% confidence intervals associated with the slope and *Y* intercept listed in the Y4 column have the largest ranges, and therefore the largest error, even though the value of the slope is the same as that obtained from fitting the perfect data set. It is also worth noting that in the Y5 column, the 95% confidence interval for the slope does not even encompass the slope of 3.0 that is obtained from fitting the other Y data sets versus the X data. Hence, introducing 20% error in the last data point results in a significant change in the slope of the line.

R^2 provides a measure of how well the data fit a line. Table 7.1 also lists the R^2 values determined from curve fitting the plots of the data in the six Y columns versus the X column. For perfect data, the R^2 value is 1. The data in the Y4 column, when plotted versus X, fit a line most poorly (lowest R^2 value), which is consistent with the large 95% confidence intervals for the slope and Y intercept obtained from this fit. Although higher R^2 values are desirable with respect to assessing how well the data fit a line, it is important to consider the 95% confidence intervals to evaluate the error in the slope and Y intercept.

Hyperbolas: Affinity Constants

Fitting data. In general, solving for variables by fitting data that are not linear is referred to as nonlinear regression, which requires specialized software. We have used Prism to perform nonlinear regression as described in the boxed instructions on page 137. Data obtained from experiments to determine an equilibrium dissociation constant (K_D) are hyperbolic and can be fit with the following equation (Chapter 2):

$$\mathrm{AB} = AB_{max}\left(\frac{[\mathrm{B}]}{[\mathrm{B}] + K_D}\right) \quad (2)$$

The measured AB values should be entered into the Y column, and the [B] values entered into the X column of Prism. Equation 2 can be entered into Prism using the following format:

$$Y = AB_{max} * X/(X + K_D) \quad (3)$$

Here, Y represents the amount of AB, X represents the concentration of B, and "*" is Prism's designation for multiplication. The two variables that will be obtained from the nonlinear regression are K_D and AB_{max}, the latter of which is the plateau amount of AB reached at high concentrations of B.

Initial values for the variables K_D and AB_{max} must be entered into the nonlinear regression software. The best approach to estimate the initial values is to plot the data (not fit the data) using spreadsheet software that has a scatter plotting function (e.g., Microsoft Excel). Estimate the AB value at which the curve plateaus (this is the initial value for AB_{max}) and the concentration of B at which the data reach half of the estimated AB_{max} (this is the initial value for K_D). Often, if the data fit the curve well, a very wide range of initial values can be used and the regression software will successfully determine K_D and AB_{max}.

Shown in the first two columns of Table 7.2 is a sample data set for a hypothetical experiment in which the amount of AB complex was measured at 19 different concentrations of B. Using Prism, nonlinear regression was performed on this data set using initial values for K_D and AB_{max} of 100 and 10,000, respectively. The K_D and AB_{max} values obtained from the curve

TABLE 7.2. Binding Affinity Sample Data

[B], nM	AB(sample)	AB(perfect)	AB(error)
10	991	887	887
20	1,533	1,632	1,632
30	2,423	2,266	2,266
40	3,143	2,813	2,813
50	3,833	3,289	3,289
60	3,375	3,708	3,708
70	4,200	4,078	4,078
80	3,956	4,409	4,409
90	4,547	4,705	4,705
100	5,300	4,973	4,973
150	5,100	5,996	5,996
200	6,867	6,684	6,684
250	7,857	7,178	7,178
300	7,050	7,550	7,550
400	8,400	8,073	8,073
500	8,500	8,423	8,423
700	8,313	8,862	8,862
900	9,450	9,126	9,126
1200	9,415	9,371	**7,497**
K_D, nM	104.9	104.9	91.4
95% CI	85.7–124.2		75.1–107.8
AB_{max}	10,190	10,190	9,588
95% CI	9,587–10,793		9,054–10,122
R^2	0.977	1.000	0.978

fit, as well as the 95% confidence intervals (95% CI) and R^2, are shown at the bottom of the AB(sample) column. This data set can be used as an initial test of any nonlinear regression software to ensure that the software is being used correctly. Plots of the sample data and the curves resulting from the nonlinear regression are shown in Figure 7.2. The upper plot shows data plotted with a linear *X* axis and the lower plot shows data plotted with a logarithmic *X* axis. It is important to realize that the data and curves in the two plots are identical; nonlinear regression was first performed on the data set, then the data set and resulting curve were plotted with two different *X* axes.

The higher* Y *values will dominate the fit. To consider how error in *Y* values affects the variables determined from nonlinear regression, we generated a perfect data set, consisting of the [B] and AB(perfect) columns in Table 7.2. This perfect series of AB amounts was calculated using Equation 2, the [B] values shown, and the K_D and AB_{max} listed in the

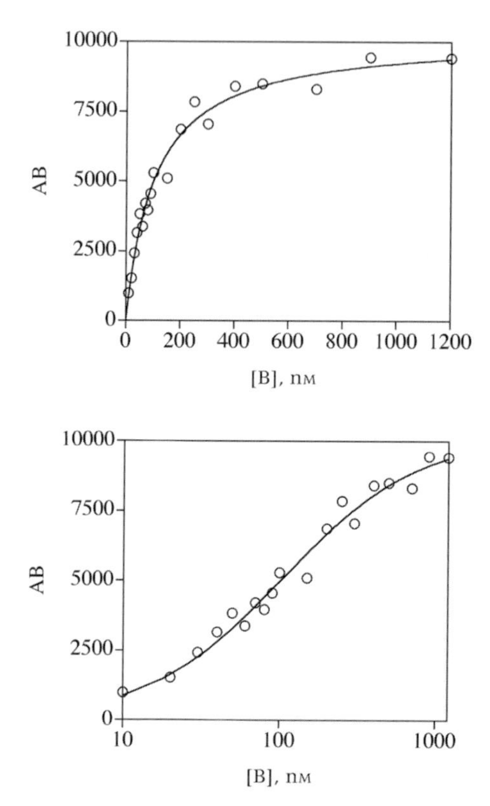

FIGURE 7.2. Sample binding data plotted on a linear *X* axis (*upper plot*) and a logarithmic *X* axis (*lower plot*). The data points are from the [B] and AB(sample) columns of Table 7.2, which were fit using Prism and Equation 3 as described. The K_D and AB_{max} values from the fit are 104.9 ± 19.2 nm and 10,190 ± 603, respectively, with errors obtained from the 95% confidence intervals.

AB(sample) column. The perfect data set can be used to demonstrate the effect of error in points in different regions of the curve on the K_D and AB_{max} values obtained from nonlinear regression. If the AB values of the perfect data points are reduced by 20% one at a time and then each data set (with one imperfect point) is plotted versus [B] and fit with Equation 3, error in the points toward the right end of the curve (i.e., the higher AB values) will have a more pronounced effect on the K_D and AB_{max}.

Consider the example shown in the AB(error) column of Table 7.2, where the AB value of the last point is decreased by 20% to 7497 (notice that all other points are the same as those in the AB[perfect] column). In this case, the values for K_D and AB_{max} decrease to 91.4 and 9588, respectively, compared to the values obtained from fitting the perfect data set. In contrast, if the AB value of the first point is decreased by 20% (to 709), the K_D and AB_{max} will change only subtly from the fit of the perfect data, to 105.4 and 10,200, respectively.

In Figure 7.3, the perfect data (AB[perfect] shown as open squares) and the AB(error) data (crosses) are plotted versus [B] along with the curves from nonlinear regression. The upper panel shows data plotted with a linear *X* axis, whereas the lower plot uses a logarithmic *X* axis. The difference in the AB_{max} (plateau) values obtained from the two different curve fits can be clearly seen. It is not possible, however, to visually detect the 13% difference in the K_D values obtained from the two data sets. Recall that we described in Chapter 2 how K_D values can be estimated by viewing the plot and determining the concentration of B at which half of

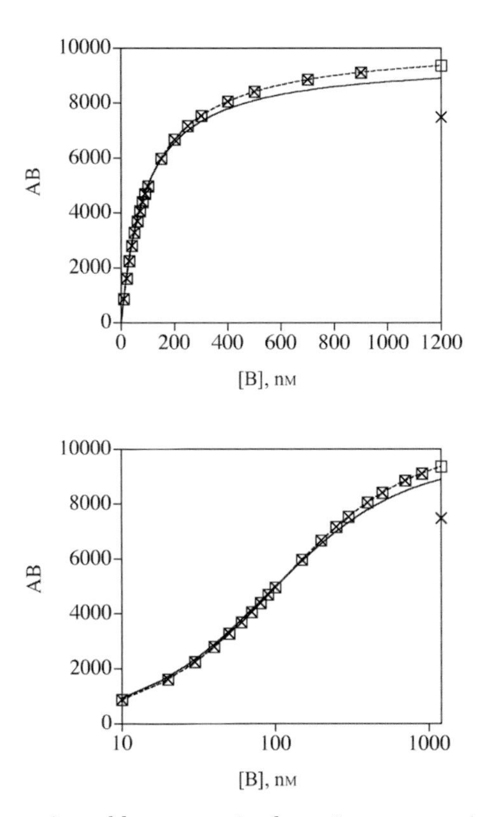

FIGURE 7.3. Illustration of how error in the point representing the highest AB value can influence the fit of the data by nonlinear regression. The *open squares* are the data set from the [B] and AB(perfect) columns of Table 7.2; the *crosses* are the data set from the [B] and AB(error) columns in Table 7.2. The lines were obtained by nonlinear regression using the AB(perfect) data (*dashed line*) and the AB(error) data (*solid line*). The *upper* and *lower panels* show the same data plotted on a linear and a logarithmic *X* axis, respectively. The K_D and AB_{max} values from the fits are shown in Table 7.2.

the maximal amount of AB forms. This approach, however, will not allow resolution of differences in K_D values between closely related data sets. Therefore, using nonlinear regression to fit affinity data sets is critical for obtaining an accurate K_D value and, in addition, enables an estimate of the error associated with the measured K_D.

Exponential Accumulations: Association Rate Constants

Fitting data. Exponential accumulation of a product is monitored when determining observed association rate constants, and the data can be fit with the following equation (Chapter 4):

$$AB_t = AB_{max}\ (1 - e^{-k_{obs}t}) \tag{4}$$

To fit these types of data using Prism, the amount of AB complex at every time point (AB_t values) are entered as Y values in Prism. The time points are entered as the X values. Equation 4 should be entered into Prism in the following format:

$$Y = AB_{max} * (1 - \exp[-k * X]) \tag{5}$$

Here, Y represents the amount of AB at each time point, X represents the time points, and "*" and "exp" are Prism's designations for multiplication and exponential, respectively. The two variables that will be obtained from the nonlinear regression are k_{obs} and AB_{max}, the latter of which is the plateau amount of AB reached at long time points.

Initial values for the variables k_{obs} and AB_{max} must be entered into the nonlinear regression software. Estimates of the initial values can be determined by plotting the data on a scatter plot using spreadsheet software. To obtain an initial value of AB_{max}, estimate the amount of AB in the plateau region of the curve. To obtain an initial value of k_{obs}, take the inverse of the time point at which the reaction is approximately 60% complete. Often, if the data fit the curve well, a very wide range of initial values can be used and the regression software will successfully determine k_{obs} and AB_{max}.

Shown in Table 7.3 is a sample data set obtained from a hypothetical experiment in which the amount of AB complex (AB[sample]) was measured at 14 different time points. Nonlinear regression was performed on this data set using Prism and initial values for k_{obs} and AB_{max} of 1 and 50,000, respectively. The k_{obs} and AB_{max} values obtained from nonlinear regression, as well as the 95% confidence intervals (95% CI) and R^2 are shown at the bottom of the AB(sample) column. This data set can be used as an initial test of any nonlinear regression software to ensure that the software is being used correctly. A plot of the sample data and curve are shown in Figure 7.4.

TABLE 7.3. Association Rate Sample Data

Time, s	AB(sample)	AB(perfect)	AB(error)
0.02	1,209	1,127	1,127
0.06	3,196	3,307	3,307
0.10	6,219	5,393	5,393
0.15	9,472	7,875	7,875
0.20	9,602	10,222	10,222
0.30	15,419	14,546	14,546
0.40	16,964	18,418	18,418
0.60	24,636	24,989	24,989
0.80	32,707	30,256	30,256
1.00	29,699	34,479	34,479
1.50	42,987	41,728	41,728
2.00	50,011	45,899	45,899
3.00	45,716	49,682	49,682
4.00	52,068	50,935	**40,748**
$k_{obs,}$ s^{-1}	1.11	1.11	1.32
95% CI	0.89–1.32		1.08–1.56
AB_{max}	51,555	51,555	46,864
95% CI	47,483–55,628		43,593–50,135
R^2	0.983	1.000	0.984

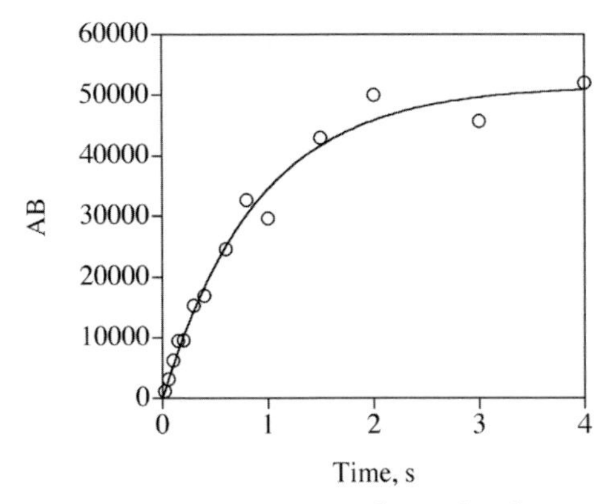

FIGURE 7.4. Plot of sample association rate data. The data points are shown in columns time and AB(sample) of Table 7.3. The curve shown was obtained by nonlinear regression. The k_{obs} and AB_{max} values from the fit are 1.11 ± 0.21 s^{-1} and 51,555 ± 4072, respectively, with errors obtained from the 95% confidence intervals.

The higher* Y *values will dominate the fit. The following analysis relies on a data set consisting of the time and AB(perfect) columns in Table 7.3, which is a perfect exponential accumulation of product over time. The series of perfect AB values were generated using Equation 4, the time points shown, and the k_{obs} and AB_{max} values in the AB(sample) column. The perfect data set can be used to demonstrate the effect of error in points in different regions of the curve on the k_{obs} and AB_{max} values obtained from nonlinear regression. If the AB values of the perfect data points are reduced by 20% (one at a time) and then each data set (with one imperfect point) is plotted versus time and fit with Equation 5, error in the points toward the right end of the curve (i.e., the higher AB values) will have a more pronounced effect on the k_{obs} and AB_{max}.

This is illustrated by the example shown in the AB(error) column of Table 7.3 in which the AB value of the last point was decreased by 20% to 40,748 compared to the AB(perfect) column (all other points are the same). When the AB(error) data are plotted versus time and fit to a curve using nonlinear regression, k_{obs} increases to 1.32 and AB_{max} decreases to 46,864 compared to the values obtained from fitting the perfect data set. Hence, the reaction appears to reach a lower plateau value sooner. In contrast, if the AB value of the first point in the AB(perfect) column is decreased by 20% (to 901), the k_{obs} and AB_{max} obtained from the curve fit will not change compared to those obtained from fitting the perfect data set. The effect of error in the last point is illustrated in Figure 7.5 in which

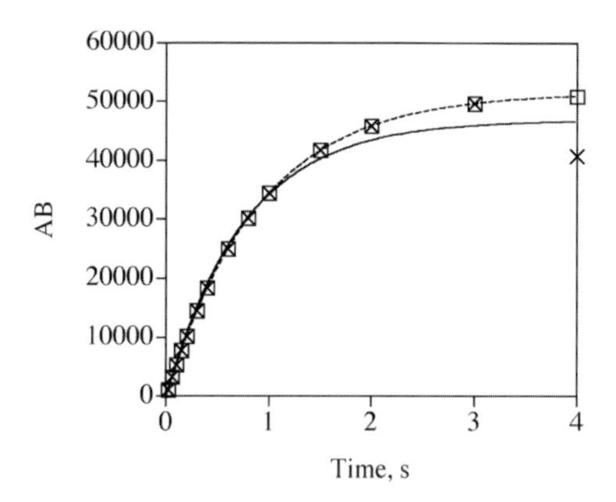

FIGURE 7.5. Illustration of how error in the point representing the highest AB value can influence the fit of exponential accumulation data by nonlinear regression. The AB(perfect) data (*open squares*) and the AB(error) data (*crosses*) were plotted versus time. Nonlinear regression curves are shown for the AB(perfect) data set (*dashed line*) and the AB(error) data set (*solid line*). The k_{obs} and AB_{max} values from the fits are shown in Table 7.3.

the AB(perfect) data, shown as open squares, and the AB(error) data, shown as crosses, are plotted versus time and analyzed by nonlinear regression. In general, adding additional data points in the plateau region of the curve will allow a more accurate value of the plateau to be established in the presence of erroneous points and likely reduce the error in the measurement of k_{obs}.

Exponential Decays: Dissociation Rate Constants

Fitting data. Exponential decay of a product is monitored when determining dissociation rate constants, which can be fit to the following equation (Chapter 4):

$$AB_t = (AB_{max} - AB_{min})(e^{-k_{-1}t}) + AB_{min} \quad (6)$$

To fit these types of data using Prism, the amount of AB complex at every time point (AB_t values) is entered as Y values in Prism. The time points are entered as the X values. Equation 6 should be entered into Prism in the following format:

$$Y = (AB_{max} - AB_{min}) * \exp(-k * X) + AB_{min} \quad (7)$$

Here, Y represents the amount of AB at each time point, X represents the time points, and "*" and "exp" are Prism's designations for multiplication and exponential, respectively. The three variables that will be obtained from the nonlinear regression are k_{-1}, AB_{max} (the amount of AB present at the zero time point determined from the fit of the data), and AB_{min} (the point at which the amount of AB levels out, reached at long time points).

Initial values for the variables k_{-1}, AB_{max}, and AB_{min} must be entered into the nonlinear regression software. Estimates of the initial values can be determined by plotting the data on a scatter plot using spreadsheet software. To obtain the initial value of AB_{max}, use the amount of AB at the zero time point (for an explanation of the zero time point, see Chapter 4). To obtain the initial value of AB_{min}, estimate the amount of AB at long times where the curve appears to level out. To estimate the initial value of k_{-1}, calculate the inverse of the time point at which the reaction is approximately 60% complete. Often, if the data fit the curve well, a very wide range of initial values can be used and the regression software will successfully determine k_{-1}, AB_{max}, and AB_{min}.

Shown in Table 7.4 is a sample data set (AB[sample]) obtained from a hypothetical decay experiment in which the amount of AB complex

TABLE 7.4. Dissociation Rate Sample Data

Time, s	AB(sample)	AB(perfect)	AB(error)
20	8,860	8,999	**7,200**
60	7,713	8,042	8,042
100	8,217	7,200	7,200
150	5,838	6,287	6,287
200	5,765	5,510	5,510
250	4,400	4,847	4,847
300	4,170	4,282	4,282
350	4,066	3,801	3,801
400	2,894	3,391	3,391
450	3,342	3,041	3,041
500	2,810	2,743	2,743
600	2,467	2,273	2,273
800	1,544	1,683	1,683
1,000	1,399	1,372	1,372
1,200	1,195	1,207	1,207
k_{-1}, s^{-1}	0.0032	0.0032	0.0025
95% CI	0.0024–0.0041		0.0018–0.0033
AB_{max}	9,526	9,526	8518
95% CI	8,751–10,301		7,878–9,159
AB_{min}	1,024	1,024	661
95% CI	272–1,776		0–1,527
R^2	0.976	1.000	0.977

was measured at 15 different time points. Nonlinear regression was performed on this data set using Prism and initial values for k_{-1}, AB_{min}, and AB_{max} of 0.005, 1,000, and 10,000, respectively. An additional constraint was used during the nonlinear regression: The value of AB_{min} was not allowed to be less than zero, because it is impossible to have a negative amount of AB complex. In Prism, this limitation is accomplished by including additional steps between Steps 10 and 11 in the boxed instructions on page 137. After completing Step 10, click on the "Default constraints" button. Under the pulldown list next to AB_{min}, select "Must be greater than," and enter 0 in the value column. Then proceed to Step 11.

The k_{-1}, AB_{max}, and AB_{min} values obtained from nonlinear regression, as well as the 95% confidence intervals (95% CI) and R^2, are shown at the bottom of the AB(sample) column. This data set can be used as an initial test of any nonlinear regression software to ensure that the software

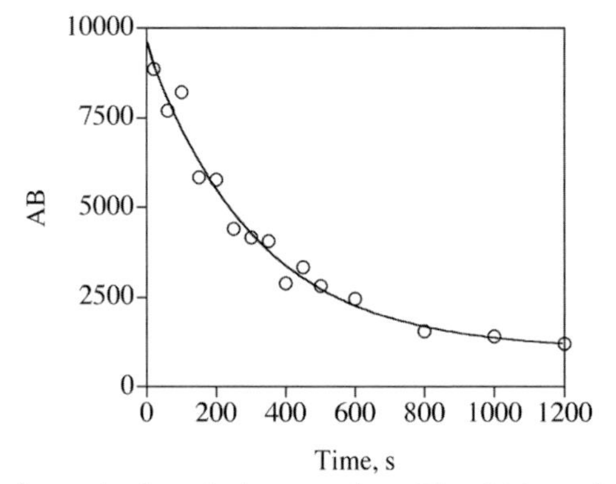

FIGURE 7.6. Plot of sample dissociation rate data. The AB(sample) data were plotted versus time. The curve shown was obtained by nonlinear regression. The k_{-1}, AB_{max}, and AB_{min} values from the fit are 0.0032 ± 0.008 s^{-1}, 9526 ± 775, and 1024 ± 752, respectively, with errors obtained from the 95% confidence intervals.

is being used correctly. A plot of the sample data and curve are shown in Figure 7.6.

The higher* Y *values will dominate the fit. The following analysis relies on a data set consisting of the time and AB(perfect) columns in Table 7.4, which is a perfect exponential decay of product over time. The series of perfect AB values was generated using Equation 6, the time points shown, and the k_{-1}, AB_{min}, and AB_{max} values in the AB(sample) column. The resulting perfect data can be used to demonstrate the effect of error in points in different regions of the curve on the k_{-1}, AB_{max}, and AB_{min} values obtained from nonlinear regression. If the AB values of the perfect data points are reduced by 20% (one at a time) and then each data set (with one imperfect point) is plotted versus time and fit with Equation 7, the error in the points toward the left end of the curve (i.e., the higher AB values at the earlier time points) have a more pronounced effect on k_{-1}, AB_{max}, and AB_{min}.

This is illustrated by the example shown in the AB(error) column of Table 7.4 in which the AB value of the first point was decreased by 20% to 7,200 compared to the AB(perfect) column (all other points are the same). When the AB(error) data are plotted versus time and curve fit using nonlinear regression, k_{-1}, AB_{max}, and AB_{min} all, respectively, decrease to 0.0025, 8,518, and 661 compared to the values shown in the AB(perfect) column. Hence, the reaction appears to level out at a lower point more slowly. In contrast, if the AB value of the last point in the AB(perfect) column is decreased by 20% (to 966) and then the data are plotted versus time and curve fit, the resulting k_{-1} and AB_{max} will be only slightly affected, whereas the AB_{min} will decrease from 1,024 to 894. In general, adding additional early time points as well as late time points to the curve will allow more accurate values of all of the variables to be determined.

ASSESSING ERROR

When making a measurement of an affinity constant or a rate constant, the investigator must assess and then report the level of confidence in the measured value. The level of confidence is most clearly expressed by stating the error associated with the experimental measurement, along with the method used to determine the error. For example, an affinity constant could be reported as 12.0 ± 1.5 nM, with the error representing the 95% confidence interval from the nonlinear regression.

Accuracy, Precision, and Error

To truly appreciate the implications of error, two terms must be understood: accuracy and precision. To better envision the meaning of and understand the distinction between accuracy and precision, consider the following illustration about four individuals playing darts.

Illustration 7.1. Julie, Max, Cristen, and Leah play darts.

Julie and Max are throwing darts at a dartboard that has a small bull's-eye in the middle. Julie grabs five red darts, aims for the bull's-eye, throws the first dart, and hits the bull's-eye dead center. This throw was highly accurate (i.e., the dart hit the intended target). Julie then throws the four remaining red darts and hits the bull's-eye all four times; therefore, each of her throws was accurate. In addition, Julie threw the five darts with a high level of precision because all of the darts landed very close together. Max then grabs five green darts, aims at the dart board, and hits the bull's-eye with the first green dart. His next four throws, however, miss the bull's-eye, and these green darts are scattered about the dart board. Max's first throw was accurate, his next four throws were less accurate, and overall the five throws were not made with great precision.

To determine the level of accuracy of each throw or the level of precision for each set of five throws requires measurements to be made. For example, the accuracy of each throw would be related to the distance of each dart from the center of the dartboard: The smaller the number, the more accurate the throw. The precision of each set of five throws would be determined by averaging or summing the measured distances for the five darts in a set: The smaller the number, the more precise the throws. The quantitative assessment of the level of precision is a crude method for estimating the error with which Julie and Max threw the darts. Julie had less error than Max.

The illustration can be extended to make it more relevant to measuring an affinity constant or rate constant. After Julie and Max leave, Cristen and Leah come upon the dartboard containing the five red darts that Julie threw and the five green darts that Max threw. Not being familiar with the game of darts, Cristen and Leah wonder where each of the two individuals who threw the darts was aiming. They do not presume that the target was the center of the dartboard, but instead decide to measure the distance of each dart from a fixed point on the board. Upon averaging the measurements for the set of red darts and also for the set of green darts, Cristen and Leah conclude that the person who threw the red darts had aimed at the center of the dart board, and the person who threw the green darts had aimed 2 inches to the right of the center of the board. However, Cristen and Leah are more confident in their conclusion about the red darts than the green darts. This is because the red darts are closely clustered (thrown with more precision) and the green darts are spread over a wider area (thrown with less precision); therefore, they know that the level of error is greater for the green darts than for the red.

This is quite similar to measuring an affinity constant or a rate constant; namely, investigators do not know what the answer should be ahead of time. Instead, a value is determined from an average of multiple measurements, and the investigators rely on an assessment of the error in the measurement (often by quantitating precision) to state the confidence that they have in the accuracy of the measured constant.

Averages and Standard Deviations

Throughout this chapter, we have used 95% confidence intervals as a means of expressing the error associated with experimentally measured rate constants. These confidence intervals are obtained by considering the variance of the multiple points from the line or curve obtained from regression analysis. Another common means of determining error associated with a measurement is to make multiple measurements of the same affinity constant or rate constant and assess the precision of the measurements. In this case, an average constant can be obtained by calculating the mean of the multiple measurements, and the error can be assessed by calculating the standard deviation of these measurements. Averages and standard deviations can be calculated using spreadsheet programs such as Excel. The standard deviation will have the same units as the average. If only two measurements are made, then the calculation of the standard deviation is not relevant, and the value of the error should be stated as half of the range between the two numbers.

Whatever the method of determining the error associated with a measured value, it is important to state the value with the associated error range. For example, a K_D could be expressed as 60 ± 12 nM, where the K_D is the mean of four measurements and the error is equivalent to one standard deviation from the mean. Or, the k_{-1} is 0.0027 ± 0.0005 s^{-1}, where the k_{-1} is the mean of two numbers and the error is equal to half of the range of the two measurements (0.0022 s^{-1} and 0.0032 s^{-1}).

The reason to present errors along with measured constants is to provide an indication of the confidence that is associated with the measured value. For example, the confidence that the K_D for a binding reaction is 60 nM is higher for 60 ± 1 nM than for 60 ± 12 nM. Naturally, the smaller the error, the more confidence in the number.

Reporting a Measured Affinity or Rate Constant

As we explained above, the confidence that is placed on a value is related to the error associated with the measurement. When reporting the results of quantitative experiments, it is necessary to decide how many digits to use in stating the measured value of an affinity or rate constant so that the value does not appear to be more or less accurate than is justified by the error. A simple guideline for deciding how to state the value of a measured constant is most easily explained by an example.

Consider that the K_D for a binding reaction is measured three times and the average of the measurements obtained from the curve fitting program is 3.5872×10^{-8} M. The standard deviation of the average is 0.09132×10^{-8} M. Hence, error first appears in the second digit after the decimal point. The confidence in the measured K_D cannot be extended beyond this digit. Therefore, the K_D should be reported as $3.59 \pm 0.09 \times 10^{-8}$ M. Once the number of digits to report has been determined, the numbers can be taken out of scientific notation if desired: $K_D = 35.9 \pm 0.9$ nM. If the error associated with this measured K_D was larger (e.g., if the standard deviation was 0.4589×10^{-8} M), then the K_D would be reported as $3.6 \pm 0.5 \times 10^{-8}$ M or 36 ± 5 nM.

As illustrated by this example, the value of the measured constant should be reported to include the digit after the decimal point at which the error first appears. It is simplest to do so when (1) the measured value for the constant is written in scientific notation with a single nonzero digit before the decimal point and (2) the error is written in scientific notation to the same power of ten as the constant.

Are the Measured Values Reasonable?

Clearly, determining the error in a measurement is extremely important; however, it does not provide an assessment of whether or not the value of a measured affinity or rate constant is reasonable and realistic. This must be assessed by the investigators, taking into account the additional knowledge that they have concerning the biomolecules being studied and the experimental system being used. Even highly accurate measurements made with great precision might not provide a reasonable measurement of an affinity or rate constant when considered in light of the biological system. For example, consider the case in which five affinity measurements, for a protein binding a specific DNA site, were made over a few days to arrive at a K_D of 20 ± 2 μM. In a previous study, however, the affinity of this protein for a related DNA sequence was estimated to be in the low nanomolar range. Should the 20 μM number be trusted? The error is low, so the measurements were precise and the confidence in the accuracy of the measured K_D value was fairly high. But, the number just does not seem reasonable with respect to the earlier estimates of the K_D (at a related site). In this case, it would be prudent not to trust the newly measured value immediately but, instead, to consider possible sources of problems with the experimental system. Perhaps the pH of the buffer or the concentration of a stock solution was too low or too high, or perhaps the fractional activity of the protein preparation declined significantly during storage in the freezer. Regardless of the reason, it is essential that the investigators continue to think about the data and the results, even when using a computer program to fit the data and estimate error.

APPENDIX 1

Equation Derivations

In the pages that follow, we show derivations for some of the more important equations contained in this volume. It is not necessary to understand how an equation is derived to use the equation; however, a deeper understanding of an equation and its meaning can often be obtained by considering the derivation. For reference, equations that are boxed in chapters are also boxed here and referred to by their chapter and equation numbers. Note that in the derivations, some "rearrange" steps result from multiple algebraic rearrangements.

Derivations covered:

- Equations for affinity constants
- Equations for competition
- Equations for homodimerization
- The Hill equation
- Equations for dissociation rate constants
- Equations for association rate constants
- Equations for half-times
- Equations for a two-step binding reaction

EQUATIONS FOR AFFINITY CONSTANTS

The model for a bimolecular binding reaction is

$$\mathrm{A} + \mathrm{B} \rightleftharpoons \mathrm{AB}$$

The total concentration of A ($[\mathrm{A}]_\mathrm{T}$, where T stands for total) is split between free A and A in the AB complex.

$$[\mathrm{A}]_\mathrm{T} = [\mathrm{A}] + [\mathrm{AB}]$$

Rearrange:

$$[\mathrm{A}] = [\mathrm{A}]_\mathrm{T} - [\mathrm{AB}]$$

The K_D for the reaction relates to the concentrations of A, B, and AB at equilibrium:

$$K_\mathrm{D} = \frac{[\mathrm{A}][\mathrm{B}]}{[\mathrm{AB}]}$$

Rearrange:

$$K_D[AB] = [A][B]$$

Substitute for A:

$$K_D[AB] = [A]_T[B] - [AB][B]$$

Rearrange:

$$[AB](K_D + [B]) = [A]_T[B]$$

Rearrange:

$$\frac{[AB]}{[A]_{Total}} = \frac{[B]}{[B] + K_D}$$

Chapter 2 Equation 6

(See also Chapter 3, Equations 7 and 8.)

For some experiments, the fraction of A bound by B may not reach a plateau of 1, often due to a portion of the A molecules being inactive. Data can be fit with an equation containing a variable f_{max} that normalizes the entire curve such that it does not have to plateau at 1.

$$\frac{[AB]}{[A]_{Total}} = f_{max}\left(\frac{[B]}{[B]+K_D}\right)$$

Chapter 2 Equation 8

It is also possible to fit data when the amount of AB is measured directly (as opposed to the fraction of A that is in the AB complex). The equation to use in this situation follows directly from the previous equation. $[AB]_{max}/[A]_{Total}$ is substituted for f_{max}. The $[A]_{Total}$ values on both sides cancel each other, and [AB] and $[AB]_{max}$ are replaced with AB and AB_{max} (which are amounts, not concentrations) to give the following equation:

$$AB = AB_{max}\left(\frac{[B]}{[B] + K_D}\right)$$

Chapter 2 Equation 9

(See also Chapter 3, Equations 10 and 11.)

This equation can be modified for use when performing an enzymatic assay to monitor the association between a regulator (R) and an enzyme (E). In this case, AB is replaced with E + ER because both free E and the ER complex might be able to synthesize the product. The term E_i is added to the right side of the equation because E might have activity alone. Finally, AB_{max} is replaced with $(ER_{max} - E_i)$, because this represents the maximum change in enzymatic activity caused by the binding of R to E.

$$E + ER = E_i + (ER_{max} - E_i)\left(\frac{[R]}{[R] + K_D}\right)$$

Chapter 2 Equation 25

EQUATIONS FOR COMPETITION

The model for competition between two molecules for binding a third is

$$A + B + X \underset{K_{D(AX)}}{\overset{K_{D(AB)}}{\rightleftharpoons}} \begin{matrix} AB \\ AX \end{matrix}$$

The total concentration of A ($[A]_T$, where T stands for total) is split between free A and A in the AB and AX complexes:

$$[A]_T = [A] + [AB] + [AX]$$

Rearrange:

$$[AB] = [A]_T - [A] - [AX]$$

Expressions for the two K_D values are:

$$K_{D(AB)} = \frac{[A][B]}{[AB]} \qquad K_{D(AX)} = \frac{[A][X]}{[AX]}$$

Substitute for [AB] in the equation for $K_{D(AB)}$:

$$K_{D[AB]} = \frac{[A][B]}{[A]_T - [A] - [AX]}$$

Rearrange:

$$[A] = \frac{K_{D[AB]}[A]_T - K_{D[AB]}[AX]}{[B] + K_{D[AB]}}$$

Use this expression to substitute for A in the equation for $K_{D(AX)}$:

$$K_{D(AX)} = \frac{K_{D(AB)}[A]_T[X] - K_{D(AB)}[AX][X]}{[B][AX] + K_{D(AB)}[AX]}$$

Rearrange:

$$\frac{[AX]}{[A]_T} = \frac{K_{D(AB)}[X]}{K_{D(AX)}[B] + K_{D(AX)}K_{D(AB)} + K_{D(AB)}[X]}$$

Divide terms on the right side by $K_{D(AB)}$ and rearrange:

$$\frac{[AX]}{[A]_T} = \frac{[X]}{[X] + K_{D(AX)}\left(1 + \frac{[B]}{K_{D(AB)}}\right)}$$

IC_{50} = [X] when $[AX]/[A]_T = 0.5$. Therefore, when $[AX]/[A]_T = 0.5$,

$$[X] = K_{D(AX)}\left(1 + \frac{[B]}{K_{D(AB)}}\right) = IC_{50}$$

Under conditions where $[B] >> K_{D(AB)}$, the above equation simplifies to:

$$IC_{50} = \frac{K_{D(AX)}[B]}{K_{D(AB)}}$$

Rearrange:

$$\boxed{\frac{K_{D(AX)}}{K_{D(AB)}} = \frac{IC_{50}}{[B]}}$$

Chapter 2 Equation 15

Moreover, when $[B] >> K_{D(AB)}$, free [A] = 0. Therefore,

$$[AB] = [A]_T - [AX]$$

Divide both sides of the equation by $[A]_T$:

$$\frac{[AB]}{[A]_T} = 1 - \frac{[AX]}{[A]_T}$$

Substitute for $[AX]/[A]_T$:

$$\frac{[AB]}{[A]_T} = 1 - \frac{[X]}{[X] + K_{D(AX)}\left(1 + \frac{[B]}{K_{D(AB)}}\right)}$$

Replace term in the denominator with IC_{50}:

$$\boxed{\frac{[AB]}{[A]_{Total}} = 1 - \frac{[X]}{[X] + IC_{50}}}$$

Chapter 2 Equation 11

EQUATIONS FOR HOMODIMERIZATION

The model for homodimerization is

$$A + A \rightleftharpoons AA$$

The K_D for the reaction relates to the concentrations of monomeric and dimeric A at equilibrium:

$$K_D = \frac{[A]^2}{[AA]}$$

Rearrange:

$$[AA] = \frac{[A]^2}{K_D}$$

The total concentration of A ($[A]_T$, where T stands for total) is split between free A and dimeric A:

$$[A]_T = [A] + 2[AA]$$

Substitute for [AA]:

$$[A]_T = [A] + \frac{2[A]^2}{K_D}$$

Rearrange:

$$2[A]^2 + K_D[A] - K_D[A]_T = 0$$

This is of the form:

$$ax^2 + bx - c = 0$$

With solution:

$$x = \frac{-b \pm \sqrt{b^2 + 4ac}}{2a}$$

Therefore,

$$[A] = \frac{-K_D \pm \sqrt{K_D^2 + 8K_D[A]_T}}{4}$$

Recall that:

$$[A] = [A]_T - 2[AA]$$

Substitute for [A] and rearrange:

$$[A]_T = \frac{-K_D + \sqrt{K_D^2 + 8K_D[A]_T}}{4} + 2[AA]$$

Divide both sides of the equation by $[A]_T$:

$$1 = \frac{-K_D + \sqrt{K_D^2 + 8K_D[A]_T}}{4[A]_T} + \frac{2[AA]}{[A]_T}$$

Rearrange:

$$\boxed{\frac{[AA]}{[A]_{Total}} = \frac{1}{2} + \frac{-K_D - \sqrt{K_D^2 + 8K_D[A]_{Total}}}{8[A]_{Total}}}$$

Chapter 2 Equation 20

THE HILL EQUATION

The model for the association between A and *n* molecules of B that is the basis for the Hill equation is:

$$A + nB \rightleftharpoons AB_n$$

This assumes perfect positive cooperativity, such that no AB complexes containing fewer than *n* molecules of B ever form. The K_D for the reaction relates to the concentrations of A, B, and AB at equilibrium:

$$K_D = \frac{[A][B]^n}{[AB_n]}$$

Convert to a more useful equation (as derived in the Equations for Affinity Constants section):

$$\frac{[AB_n]}{[A]_{Total}} = \frac{[B]^n}{[B]^n + K_D}$$

Invert both sides of the equation:

$$\frac{[A]_{Total}}{[AB_n]} = \frac{[B]^n + K_D}{[B]^n}$$

Rearrange:

$$\frac{[A]_{Total} - [AB_n]}{[AB_n]} = \frac{K_D}{[B]^n}$$

Invert both sides of the equation:

$$\frac{[AB_n]}{[A]_{Total} - [AB_n]} = \frac{[B]^n}{K_D}$$

Take the $\log_{10}$ of both sides of the equation:

$$\log\left(\frac{[AB_n]}{[A]_{Total} - [AB_n]}\right) = n(\log[B]) - \log(K_D)$$

Divide the numerator and denominator inside the log on the left side of the equation by $[A]_T$. Substitute Y for $[AB_n]/[A]_T$ and n_H for n:

$$\boxed{\log\left(\frac{Y}{1 - Y}\right) = n_H(\log[B]) - \log(K_D)}$$

Chapter 3
Equation 16

EQUATIONS FOR DISSOCIATION RATE CONSTANTS

The model for a biomolecular binding reaction, including the two rate constants, is:

$$A + B \underset{k_{-1}}{\overset{k_1}{\rightleftharpoons}} AB$$

The rate equation describing the change in AB over time is:

$$d[AB]/dt = k_1[A][B] - k_{-1}[AB]$$

When measuring the dissociation rate constant by blocking new association [A], or [B] is equal to zero, the previous equation therefore simplifies to:

$$d[AB]/dt = -k_{-1}[AB]$$

Rearrange:

$$d[AB]/[AB] = -k_{-1}dt$$

Take the integral:

$$\int_{[AB]_0}^{[AB]_t} d[AB]/[AB] = \int_0^t -k_{-1}\,dt$$

With solution:

$$\ln[AB]_t - \ln[AB]_0 = -k_{-1}t$$

Rearrange:

$$\ln([AB]_t/[AB]_0) = -k_{-1}t$$

Convert from natural log to exponential (where $[AB]_0 = [AB]_i$):

$$\frac{[AB]_t}{[AB]_i} = e^{-k_{-1}t}$$

Chapter 4
Equation 13

Multiply both sides of the equation by $[AB]_i$ and change from concentration of AB to amount:

$$AB_t = AB_i(e^{-k_{-1}t})$$

Chapter 4
Equation 15

For some experiments, the fraction of A bound by B at the first time point might be significantly less than 1, due to the presence of rapidly dissociating nonspecific complexes. In addition, it is possible that the data will level out above 0 at long time points. The following equation contains two additional variables: f_{max}, which is the fraction bound at zero time when the curve is extrapolated back to $t = 0$, and f_{min}, which is the minimum fraction bound at long times. The term $(f_{max} - f_{min})$ normalizes the entire curve to the maximum change in fraction bound. Finally, because the fraction bound could plateau at a point above 0, the term f_{min} is added to the right side of the equation to give the following:

$$\frac{[AB]_t}{[AB]_i} = (f_{max} - f_{min})\,(e^{-k_{-1}t}) + f_{min}$$

Chapter 4
Equation 19

It is also possible to fit data when the amount of AB is measured directly (as opposed to the fraction of A that is in the AB complex). The equation to use in this situation follows directly from the previous equation: Replace $[AB]_t/[AB]_i$ with AB_t, f_{max} with AB_{max}, and f_{min} with AB_{min}.

$$AB_t = (AB_{max} - AB_{min})\,(e^{-k_{-1}t}) + AB_{min}$$

Chapter 4
Equation 20

(See also Chapter 5, Equations 20 and 21.)

EQUATIONS FOR ASSOCIATION RATE CONSTANTS

The model for bimolecular binding reactions, including the two rate constants, is:

$$A + B \underset{k_{-1}}{\overset{k_1}{\rightleftharpoons}} AB$$

The total concentration of A ($[A]_T$, where T stands for total) is split between free A and A in the AB complex:

$$[A]_T = [A] + [AB]$$

Rearrange:

$$[A] = [A]_T - [AB]$$

The rate equation describing the change in AB over time is:

$$d[AB]/dt = k_1[A][B] - k_{-1}[AB]$$

Substitute for A and rearrange:

$$d[AB]/dt = k_1([A]_T - [AB])[B] - k_{-1}[AB]$$

Rearrange:

$$d[AB]/dt + (k_1[B] + k_{-1})[AB] - k_1[A]_T[B] = 0$$

This is of the form:

$$dx/dt + \alpha x - \beta = 0$$

With solution:

$$x = -\left(\frac{\beta}{\alpha}\right)e^{-\alpha t} + \frac{\beta}{\alpha}$$

Rearrange:

$$x = -\left(\frac{\beta}{\alpha}\right)\left(1 - e^{-\alpha t}\right)$$

Therefore,

$$[AB] = \left(\frac{k_1[A]_T[B]}{k_1[B] + k_{-1}}\right)\left(1 - e^{-(k_1[B] + k_{-1})t}\right)$$

As explained in the next section (Solving for $[AB]_{max}$), the term in the parentheses immediately to the right of the equal sign is equivalent to $[AB]_{max}$. Therefore,

$$[\mathrm{AB}] = [AB]_{max}\left(1 - e^{-(k_1[B] + k_{-1})t}\right)$$

Rearrange (and replace the [AB] with $[AB]_t$):

$$\frac{[\mathrm{AB}]_t}{[AB]_{max}} = 1 - e^{-k_{obs}t}$$

Chapter 4 Equation 24

where (under conditions where [B] >> [A], $[B] \approx [B]_i$):

$$k_{obs} = k_1[\mathrm{B}]_i + k_{-1}$$

Chapter 4 Equation 25

(See also Chapter 5, Equations 18 and 19.)

Multiply both sides of the equation by $[AB]_{max}$ and change concentrations to amounts:

$$\mathrm{AB} = AB_{max}(1 - e^{-k_{obs}t})$$

Chapter 4 Equation 27

(See also Chapter 5, Equations 16 and 17.)

Solving for $[AB]_{max}$

The maximum amount of AB that will form at long times (as the reaction approaches equilibrium) when measuring association rate constants is governed by the equilibrium constant for the reaction. At equilibrium, there is no change in [AB] over time. Hence,

$$d[\mathrm{AB}]/dt = 0$$

Therefore (see derivation above),

$$k_1[\mathrm{A}]_T[\mathrm{B}] - k_1[\mathrm{AB}][\mathrm{B}] - k_{-1}[\mathrm{AB}] = 0$$

Rearrange:

$$k_1[\mathrm{AB}][\mathrm{B}] + k_{-1}[\mathrm{AB}] = k_1[\mathrm{A}]_T[\mathrm{B}]$$

This equation can be rearranged in terms of [AB] at equilibrium, to obtain the maximal [AB] formed at the plateau of the rate curve.

$$[AB]_{max} = \frac{k_1[\mathrm{A}]_T[\mathrm{B}]}{k_1[\mathrm{B}] + k_{-1}}$$

EQUATIONS FOR HALF-TIMES

$t_{1/2}$ for Dissociation

The half-time ($t_{1/2}$) for a dissociation reaction is the time that it takes for a reaction to proceed halfway from the original bound state to the fully dissociated state after blocking new binding. Begin with the general equation for determining a dissociation rate constant.

$$\frac{[AB]_t}{[AB]_i} = e^{-k_{-1}t}$$

Set $[AB]_t/[AB]_i = 0.5$:

$$0.5 = e^{-k_{-1}t_{1/2}}$$

Take the natural log of both sides of the equation:

$$\ln(0.5) = -k_{-1}t_{1/2}$$

Multiple both sides of the equation by –1:

$$-\ln(0.5) = k_{-1}t_{1/2}$$

Because the –ln(0.5) = ln(2),

$$\ln(2) = k_{-1}t_{1/2}$$

Rearrange:

$$\boxed{t_{1/2} = \frac{\ln 2}{k_{-1}} = \frac{0.693}{k_{-1}}}$$

Chapter 4
Equation 14

$t_{1/2}$ for Association

The half-time ($t_{1/2}$) for association is the time that it takes for a reaction to proceed halfway from unbound to bound after mixing the two biomolecules. Begin with the general equation used for determining an association rate constant.

$$\frac{[AB]_t}{[AB]_{max}} = 1 - e^{-k_{obs}t}$$

Set $[AB]_t/[AB]_f = 0.5$:

$$0.5 = e^{-k_{-obs}t_{1/2}}$$

The remainder of the derivation is identical to that shown in the $t_{1/2}$ for Dissociation section, resulting in

$$\boxed{t_{1/2} = \frac{\ln 2}{k_{obs}}}$$

Chapter 4
Equation 26

EQUATIONS FOR A TWO-STEP BINDING REACTION

Rate Constant for Association

The model for a two-step biomolecular binding reaction, including the four microscopic rate constants, is

$$\mathrm{A} + \mathrm{B} \underset{k_{-1}}{\overset{k_1}{\rightleftharpoons}} \mathrm{AB} \underset{k_{-2}}{\overset{k_2}{\rightleftharpoons}} \mathrm{AB}^*$$

The concentration of free A (where T stands for total) is:

$$[\mathrm{A}] = [\mathrm{A}]_{\mathrm{T}} - [\mathrm{AB}] - [\mathrm{AB}^*]$$

Applying the assumption that the initial binding step is in rapid equilibrium compared to the equilibrium of the second transformation step (i.e., $k_1[\mathrm{B}]$, $k_{-1} >> k_2$, k_{-2})

$$K_{\mathrm{D1}} = \frac{[\mathrm{A}][\mathrm{B}]}{[\mathrm{AB}]}$$

Substituting for [A],

$$K_{\mathrm{D1}} = \frac{[\mathrm{A}]_{\mathrm{T}}[\mathrm{B}] - [\mathrm{AB}][\mathrm{B}] - [\mathrm{AB}^*][\mathrm{B}]}{[\mathrm{AB}]}$$

Rearrange:

$$K_{\mathrm{D1}} = \frac{[\mathrm{A}]_{\mathrm{T}}[\mathrm{B}] - [\mathrm{AB}^*][\mathrm{B}]}{[\mathrm{AB}]} - [\mathrm{B}]$$

Rearrange:

$$[\mathrm{AB}] = \frac{[\mathrm{A}]_{\mathrm{T}}[\mathrm{B}] - [\mathrm{AB}^*][\mathrm{B}]}{K_{\mathrm{D1}} + [\mathrm{AB}]}$$

The rate equation describing the change in AB* over time is:

$$d[\mathrm{AB}^*]/dt = k_2[\mathrm{AB}] - k_{-2}[\mathrm{AB}^*]$$

Substitute for [AB] and rearrange:

$$d[\mathrm{AB}^*]/dt + \left(\frac{k_2[\mathrm{B}] + k_{-2}K_{\mathrm{D1}} + k_{-2}[\mathrm{B}]}{K_{\mathrm{D1}} + [\mathrm{B}]}\right)[\mathrm{AB}^*] - \frac{k_2[\mathrm{A}]_{\mathrm{T}}[\mathrm{B}]}{K_{\mathrm{D1}} + [\mathrm{B}]} = 0$$

This is of the form:

$$dx/dt + \alpha x - \beta = 0$$

With solution:

$$x = \left(\frac{\beta}{\alpha}\right) e^{-\alpha t} + \frac{\beta}{\alpha}$$

Rearrange:

$$x = \left(\frac{\beta}{\alpha}\right)\left(1 - e^{-\alpha t}\right)$$

In this equation,

$$\alpha = k_{\mathrm{obs}} \qquad \frac{\beta}{\alpha} = [AB^*]_{\mathrm{max}} \qquad x = [\mathrm{AB}^*]$$

Therefore,

$$[\mathrm{AB}^*] = [AB^*]_{\mathrm{max}}(1 - e^{-k_{\mathrm{obs}}t})$$

where

$$k_{\mathrm{obs}} = \frac{k_2[\mathrm{B}] + k_{-2}K_{\mathrm{D1}} + k_{-2}[\mathrm{B}]}{K_{\mathrm{D1}} + [\mathrm{B}]}$$

Rearrange:

$$k_{\mathrm{obs}} = \frac{k_2[\mathrm{B}]}{K_{\mathrm{D1}} + [\mathrm{B}]} + k_{-2}$$

Substitute $[\mathrm{B}]_i$ for [B] (under conditions where [B] >> [A]) and eliminate k_{-2} (under conditions where $k_2 >> k_{-2}$):

$$k_{\mathrm{obs}} = \frac{k_2[\mathrm{B}]_{\mathrm{i}}}{K_{\mathrm{D1}} + [\mathrm{B}]_{\mathrm{i}}}$$

Multiply terms in the numerator and denominator by k_1:

$$k_{obs} = \frac{k_1 k_2 [B]_i}{k_1 [B]_i + k_{-1}}$$

Chapter 6
Equation 5

Double Reciprocal Plot

Invert both sides of the boxed equation for k_{obs} (from the Rate Constant for Association section):

$$\frac{1}{k_{obs}} = \frac{k_1 [B]_i + k_{-1}}{k_1 k_2 [B]_i}$$

Rearrange:

$$\frac{1}{k_{obs}} = \left(\frac{k_{-1}}{k_1 k_2}\right)\frac{1}{[B]_i} + \frac{1}{k_2}$$

Chapter 6
Equation 6

Rate Constant for Decay

Begin with the following rate equation (from the Rate Constant for Association section):

$$d[AB^*]/dt + \left(\frac{k_2[B] + k_{-2}K_{D1} + k_{-2}[B]}{K_{D1} + [B]}\right)[AB^*] - \frac{k_2[A]_T[B]}{K_{D1} + [B]} = 0$$

When measuring dissociation rate constants by blocking new association [B] = 0, the equation therefore simplifies to:

$$d[AB^*]/dt + k_{-2}[AB^*] = 0$$

Rearrange:

$$d[AB^*]/[AB^*] = k_{-2}dt$$

The remainder of the derivation is identical to that shown for k_{-1} in the Equations for Dissociation Rate Constants section:

$$\frac{[AB]_t}{[AB]_i} = e^{-k_{off}t}$$

Here,

$$k_{off} = k_{-2}$$

APPENDIX 2

Resources and References

THIS APPENDIX CONTAINS AN ANNOTATED LIST of resources and references organized by chapter that are suggestions for further reading. Included are books, examples from the literature, and Internet resources. In general, the examples from the literature experimentally illustrate the concepts we discuss in each chapter, and the books typically delve deeper into the theory and equations we present. This is by no means an all-inclusive list of resources, but rather it provides a starting point for further understanding the topics covered in this book.

CHAPTER 1

Books

Boyer R. 2000. *Modern experimental biochemistry,* 3rd Edition. Benjamin Cummings, San Francisco.

This book contains descriptions of theory and experimental techniques (as well as actual sample experiments and procedures) that are useful for quantitating biomolecules and monitoring binding reactions.

Boyer R. 2006. *Biochemistry laboratory: Modern theory and techniques.* Benjamin Cummings, San Francisco.

Discussed are theoretical and procedural aspects for designing experiments using biological molecules. In addition, this book contains information on useful Web sites for biochemists and molecular biologists, as well as useful software programs for data analysis.

Robyt J.F. and White B.J. 1990. *Biochemical techniques: Theory and practice.* Waveland Press, Prospect Heights, Illinois.

Explained here are the theories behind some of the techniques we describe for monitoring binding reactions, for example, electrophoresis, chromatography, and spectroscopic methods. In addition, the book covers methods for visualizing and quantitating biological molecules.

Journal Articles

Chmiel N.H., Rio D.C., and Doudna J.A. 2006. Distinct contributions of KH domains to substrate binding affinity of *Drosophila* P-element somatic inhibitor protein. *RNA* **12:** 283–291.

This paper contains an example of determining fractional activities for a series of mutant proteins.

Vishwanath B.S., Frey F.J., Bradbury M., Dallman M.F., and Frey B. 1992. Adrenalectomy decreases lipocortin-I messenger ribonucleic acid and tissue protein content in rats. *Endocrinology* **130:** 585–591.

The authors used quantitative western blotting to compare levels of a protein in extracts.

CHAPTER 2

Books

Cantor C.R. and Schimmel P.R. 1980. *Biophysical chemistry, Part III: The behavior of biological macromolecules.* W.H. Freeman, New York.

This is the third volume in a three-volume series covering topics in biophysics using a mathematical and theoretical approach. Chapters 15 and 23 in this book most directly apply to measuring binding affinities.

van Holde K.E., Johnson W.C., and Ho P.S. 2006. *Principles of physical biochemistry,* 2nd Edition. Pearson Prentice Hall, New Jersey.

Chapter 14 in this book covers binding equilibrium and presents equations relating binding affinity to free energy. Chapter 2 provides background on free energy and thermodynamics.

Journal Articles

Allenby G., Janocha R., Kazmer S., Speck J., Grippo J.F., and Levin A.A. 1994. Binding of 9-*cis*-retinoic acid and all-*trans*-retinoic acid to retinoic acid receptors α, β, and γ. *J. Biol. Chem.* **269:** 16689–16695.

This study used competition binding experiments to measure IC_{50} values for ligand-receptor interactions.

Battle D.J. and Doudna J.A. 2001. The stem-loop binding protein forms a highly stable and specific complex with the 3′ stem-loop of histone mRNAs. *RNA* **7:** 123–132.

See Literature Example 2.1.

Chapman-Smith A., Lutwyche J.K., and Whitelaw M.L. 2004. Contribution of the Per/Arnt/Sim (PAS) domains to DNA binding by the basic helix-loop-helix PAS transcriptional regulators. *J. Biol. Chem.* **279:** 5353–5362.

This paper contains examples of determining K_D values for protein–DNA interactions using electromobility shift assays.

Clark K.D., Garczynski S.F., Arora A., Crim J.W., and Strand M.R. 2004. Specific residues in plasmatocyte-spreading peptide are required for receptor binding and functional antagonism of insect immune cells. *J. Biol. Chem.* **279:** 33246–33252.

See Literature Example 2.2.

Coleman R.A., Taggart A.K.P., Benjamin L.R., and Pugh B.F. 1995. Dimerization of the TATA binding protein. *J. Biol. Chem.* **270:** 13842–13849.

The authors measured the K_D values for homodimerization of a transcription factor off of DNA. The paper also contains the derivation of the relevant equation.

Wild M.K., Huang M., Schulze-Horsel U., van der Merwe P.A., and Vestweber D. 2001. Affinity, kinetics, and thermodynamics of E-selectin binding to E-selectin ligand. *J. Biol. Chem.* **276:** 31602–31612.

The authors used surface plasmon resonance to measure the affinity, kinetics, and thermodynamics of a protein–protein interaction. This paper illustrates how these three quantitative assessments of biological binding reactions relate to one another.

Yuan T. and Vogel H.J. 1999. Substitution of the methionine residues of calmodulin with the unnatural amino acid analogs ethionine and norleucine: Biochemical and spectroscopic studies. *Protein Sci.* **8:** 113–121.

See Literature Example 2.3.

CHAPTER 3

Books

Ben-Naim A.Y. 2001. *Cooperativity and regulation in biochemical processes.* Kluwer Academic/Plenum, New York.

This book is devoted to the mechanisms, equations, theory, and statistical mechanics governing cooperativity.

Voet D. and Voet J.G. 1995. *Biochemistry,* 2nd Edition. John Wiley & Sons, Somerset, New Jersey.

Chapter 9 of this textbook covers quantitative analysis of oxygen binding to hemoglobin, a classic example of cooperative binding.

Journal Articles

Bruck I., Woodgate R., McEntee K., and Goodman M.F. 1996. Purification of a soluble UmuD′C complex from *Escherichia coli.* Cooperative binding of UmuD′C to single-stranded DNA. *J. Biol. Chem.* **271:** 10767–10774.

See Literature Example 3.1.

Jensen P.Y., Bonander N., Moller L.B., and Farver O. 1999. Cooperative binding of copper(I) to the metal binding domains in Menkes disease protein. *Biochem. Biophys. Acta* **1434:** 103–113.

This paper illustrates the use of a Hill plot to assess the cooperative binding of copper to a protein.

Senear D.F. and Brenowitz M. 1991. Determination of binding constants for cooperative site-specific protein-DNA interactions using the gel mobility-shift assay. *J. Biol. Chem.* **266:** 13661–13671.

Described are both simulated data and experimental electromobility shift data depicting cooperativity in protein–DNA interactions. Included are theory, mathematical derivations, binding curves, and a discussion of which K_D values in cooperative systems can and cannot be quantitatively determined.

Song M. and McHenry C.S. 2001. Carboxyl-terminal domain III of the δ′ subunit of DNA polymerase III holoenzyme binds DnaX and supports cooperative DnaX complex assembly. *J. Biol. Chem.* **276:** 48709–48715.

This paper describes gel-filtration experiments used to quantitatively examine the cooperative assembly of the bacterial DNA polymerase III holoenzyme.

Weiss J.N. 1997. The Hill equation revisited: Uses and misuses. *FASEB J.* **11:** 835–841.

The author provides a mathematical and theoretical explanation of the limitations and utility of the Hill equation.

CHAPTER 4

Books

Cantor C.R. and Schimmel P.R. 1980. *Biophysical chemistry, Part III: The behavior of biological macromolecules.* W.H. Freeman, New York.

This is the third volume in a three-volume series covering topics in biophysics using a mathematical and theoretical approach. Chapter 16 most directly applies to rate equations for binding reactions.

Fersht A. 1985. *Enzyme structure and mechanism,* 2nd Edition. W.H. Freeman, New York.

Although the main focus of this book is enzyme kinetics, it contains mathematical and written explanations of kinetic theory and principles that are applicable to binding reactions.

Journal Articles

Abel R.L., Haigis M.C., Park C., and Raines R.T. 2002. Fluorescence assay for the binding of ribonuclease A to the ribonuclease inhibitor protein. *Anal. Biochem.* **306:** 100–107.

See Literature Example 4.1.

Hammond P.W. and Cech T.R. 1997. Euplotes telomerase: Evidence for limited base-pairing during primer elongation and dGTP as an effector of translocation. *Biochemistry* **37:** 5162–5172.

Multiple dissociation rate constants were determined for the interaction between telomerase and various primers. This study illustrates how measuring dissociation rate constants can reveal information about the mechanism of a reaction.

Jamieson E.R. and Lippard S.J. 2000. Stopped-flow fluorescence studies of HMG-domain protein binding to cisplatin-modified DNA. *Biochemistry* **39:** 8426–8438.

See Literature Example 4.3.

Petri V., Hsieh M., Jamison E., and Brenowitz M. 1998. DNA sequence-specific recognition by the *Saccharomyces cerevisiae* "TATA" binding protein: Promoter-dependent differences in the thermodynamics and kinetics of binding. *Biochemistry* **37:** 15842–15849.

This study determined association rate constants for a transcription factor binding to different promoter DNA sequences. It also relates kinetic and thermodynamic parameters.

Recht M.I. and Williamson J.R. 2001. Central domain assembly: Thermodynamics and kinetics of S6 and S18 binding to an S15-RNA complex. *J. Mol. Biol.* **313:** 35–48.

See Literature Example 4.2.

Wickiser J.K., Cheah M.T., Breaker R.R., and Crothers D.M. 2005. The kinetics of ligand binding by an adenine-sending riboswitch. *Biochemistry* **44:** 13404–13414.

This study investigated the kinetics of ligand binding by a riboswitch. Shown are examples of determining both association and dissociation rate constants.

CHAPTER 5

Journal Articles

Bassi G.S. and Weeks K.M. 2003. Kinetic and thermodynamic framework for assembly of the six-component bI3 group I intron ribonucleoprotein catalyst. *Biochemistry* **42:** 9980–9988.

This paper describes differences in association and dissociation rate constants for assembly of an RNA–protein complex observed in the absence and presence of magnesium.

Cik M., Masure S., Lesage A.S.J., Van der Linden I., Van Gompel P., Pangalos M.N., Gordon R.D., and Leysen J.E. 2000. Binding of GDNF and Neuturin to human GDNF family receptor α 1 and 2. *J. Biol. Chem.* **275:** 27505–27512.

This study investigated how a protein factor affects the rate constants for association and dissociation of a ligand-receptor pair, leading to positive cooperativity. Equilibrium dissociation constants were also measured.

Gao D. and McHenry C.S. 2001. τ binds and organizes *Escherichia coli* replication proteins through distinct domains. *J. Biol. Chem.* **276:** 4447– 4453.

The authors used surface plasmon resonance to examine the kinetics of interactions between proteins involved in bacterial replication. Positive cooperativity occurred primarily through decreases in dissociation rate constants.

Nalefski E.A. and Newton A.C. 2001. Membrane binding kinetics of protein kinase C βII mediated by the C2 domain. *Biochemistry* **40:** 13216– 13229.

See Literature Example 5.1.

CHAPTER 6

Books

Fersht A. 1985. *Enzyme structure and mechanism,* 2nd Edition. W.H. Freeman, New York.

The main focus of this book is enzyme kinetics; however, Chapter 4 provides kinetic theory and derivations of equations that relate to two-step binding reactions.

Journal Articles

Jucker F.M., Phillips R.M., McCallum S.A., and Pardi A. 2003. Role of heterogeneous free state in the formation of a specific RNA-theophylline complex. *Biochemistry* **42:** 2560–2567.

The authors performed kinetic analysis of a small-molecule RNA-binding interaction. The data reveal that the reaction is better modeled as a two-step reaction, rather than a single-step reaction (the two-step model is different from the one presented in Chapter 6).

McClure W.R. 1980. Rate-limiting steps in RNA chain initiation. *Proc. Natl. Acad. Sci.* **77:** 5634–5638.

See Literature Example 6.1.

Robblee J.P., Olivares A.O., and De La Cruz E.M. 2004. Mechanism of nucleotide binding to actomyosin VI. *J. Biol. Chem.* **279:** 38608–38617.

This study investigated the kinetics of nucleotide binding to a protein, which occurs through a two-step mechanism of the type we discuss in the book.

Ross W. and Gourse R.L. 2005. Sequence-independent upstream DNA-αCTD interactions strongly stimulate *Escherichia coli* RNA polymerase-lacUV5 promoter association. *Proc. Natl. Acad. Sci.* **102:** 291–296.

Discussed are multiple kinetic analyses of a two-step binding reaction using *E. coli* RNA polymerase as a model system; the effects of polymerase and promoter deletions on the various rate constants were determined.

Sha M., Wang Y., Xiang T., van Heerden A., Browning K.S., and Goss D.J. 1995. Interaction of wheat germ protein synthesis initiation factor eIF-(iso)4F and its subunits p28 and p86 with m^7GTP and mRNA analogues. *J. Biol. Chem.* **270:** 29904–29909.

See Literature Example 6.2.

Strickland S., Palmer G., and Massey V. 1975. Determination of dissociation constants and specific rate constants of enzyme-substrate (or protein-ligand) interactions from rapid reaction kinetic data. *J. Biol. Chem.* **250:** 4048–4052.

This paper contains theory and derivations of equations that describe two-step binding reactions of the nature we discuss in Chapter 6.

CHAPTER 7

Books

Adams D.S. 2003. *Lab math: A handbook of measurements, calculations, and other quantitative skills for use at the bench.* Cold Spring Harbor Laboratory Press, Cold Spring Harbor, New York.

This book discusses interpreting and presenting numerical data, scientific notation, and significant digits, all of which are important for data analysis.

Schlager G. 1999. *Statistics for biochemists: A primer.* Madeline Press, Princeville, Hawaii.

This book reviews statistical terms, tests, and analyses as they apply to quantitative experiments in biological systems. It also covers theory behind regression.

Web Sites

www.biokin.com

This Web site contains a free software program (DynaFit; not yet available for Macintosh computers) that will perform regression analysis of binding and kinetic data. The Web site contains tutorials and examples of experimental data.

www.curvefit.com

This Web site was designed by GraphPad Software, Inc. (the company that developed Prism software), and it contains comprehensive information regarding the theory of regression analysis, how to apply it to multiple types of biological data, and how to interpret the results of nonlinear regression. Four books that cover the aforementioned topics can be downloaded as pdf files. The book—*Fitting Models to Biological Data Using Linear and Nonlinear Regression*—is very useful. Moreover, a free trial of the Prism program can be downloaded.

www.kintek-corp.com

This Web site contains a free program (KinTekSim; not yet available for Macintosh computers) that allows users to fit experimental data to predicted mechanisms, as well as to simulate data given a reaction mechanism.

Other software programs that can perform nonlinear regression:

KaleidaGraph: www.synergy.com
SigmaPlot: www.systat.com/products/SigmaPlot/

Index